Biotechnology and International Security

Biotechnology and International Security

David Malet

ROWMAN & LITTLEFIELD
Lanham • Boulder • New York • London

Published by Rowman & Littlefield
A wholly owned subsidiary of
The Rowman & Littlefield Publishing Group, Inc.
4501 Forbes Boulevard, Suite 200, Lanham, Maryland 20706
www.rowman.com

Unit A, Whitacre Mews, 26-34 Stannary Street, London SE11 4AB,
United Kingdom

British Library Cataloguing in Publication Information Available

Library of Congress Cataloging-in-Publication Data
Names: Malet, David, 1976– author.
Title: Biotechnology and international security / David Malet.
Description: Lanham, MD : Rowman & Littlefield, [2016] | Includes bibliographical
    references and index.
Identifiers: LCCN 2016014600 (print) | LCCN 2016014818 (ebook) | ISBN
    9781442268906 (cloth : alk. paper) | ISBN 9781442268913 (electronic)
Subjects: LCSH: Biological warfare. | Biological warfare—Moral and ethical aspects. |
    Biotechnology—Moral and ethical aspects. | Security, International.
Classification: LCC UG447.8 .M28 2016 (print) | LCC UG447.8 (ebook) | DDC 358/
    .3809—dc23
LC record available at http://lccn.loc.gov/2016014600

∞ ™ The paper used in this publication meets the minimum requirements of American
National Standard for Information Sciences Permanence of Paper for Printed Library
Materials, ANSI/NISO Z39.48-1992.

Printed in the United States of America

To Erica and Olivia
May it be a better world

Contents

Acknowledgments

This book has an extended backstory, and I must first and foremost thank my editor, Marie-Claire Antoine, for enabling it to finally come to fruition through her strong support for this project and her valuable suggestions for organizing the material.

The initial research and major themes date back to 1998 and a net assessment paper for Andrew Krepinevich's class at Georgetown University's Security Studies Program. While some of the experimental Pentagon and private-sector programs of nearly twenty years ago have not eventuated, nearly all of the lines of research announced since then are still active, and some that seemed fanciful in the twentieth century, such as engineered vectors to treat medical conditions, animals controlled by cybernetics, and soldiers in exoskeletons, are already a reality.

I returned to this material for a panel, organized by the Bridging the Gap project for the 2010 International Studies Association conference, that examined the prospect of WMD in the hands of private actors. The proposal by Matthew Kroenig may have been initially tongue in cheek, but there were many interesting insights provided by participants and audience members including James Goldgeier, Robert Brown, Rachel Whitlark, Amir Stepak, Amy J. Nelson, Theresa Macphail, Tanya Ogilvie-White, and Gary Ackerman. My gratitude goes to the Bridging the Gap team, including Bruce Jentleson, Steven Weber, Ely Ratner, Naazneen Barma, and Brent Durbin, for securing this panel and for supporting policy-relevant scholarship in the study of international relations.

My conference paper attracted the interest of the late Henry Tom, who urged me to develop a manuscript for his press but sadly was later unable to serve as editor. In the course of its development I received extremely helpful feedback on manuscript and article drafts and conference paper presentations from Mark Korbitz, Margaret Kosal, Manabrata Guha, Lynn Klotz, Terry O'Sullivan, Jacqui True, Andrew Phillips, Tamir Libel, Paul Lushenko, and anonymous reviewers. I also benefitted from the assistance of David Galbreath, Donald Jacobs, David McBride, and Elizabeth Demers. Richard Kreminski gave me the opportunity to give a presentation for early feedback from the faculty in the Food for Thought series at the College of Science and Mathematics at Colorado State University–Pueblo in October 2010.

Some of the material contained in this book, particularly in chapter 3, appeared previously in my article "Captain America in International Relations: The Biotech Revolution in Military Affairs" in *Defence Studies* 15(4). I am grateful to Taylor and Francis for granting permission to reprint it here.

For the sections describing the incidents in Senator Daschle's office during 2001–2002, I am indebted to my former Capitol Hill coworkers Peter Hansen, Jill Marshall, and Aaron Fischbach for aiding my recollections, and to the FBI for providing Daschle staff alumni with the report on the Amerithrax case before it was released publicly.

There is much that we can learn about emerging developments from what is already history.

Preface

MORE THAN AN ACADEMIC INTEREST

The rental car sped out of the wilderness toward the settlement as the morning sun rose in the sky. Along the sides of the potholed highway were a number of warning markers adorned with a red X, erected by the government to show where individuals had died. The many more dead dogs along the side of the road had received no such commemoration. The car slowed as the driver entered the village, and she noted as the inhabitants came into sight that health problems were rife among them, as was unemployment, which was above 90 percent. Clearly we were in a disaster zone.

It was, in short, a normal day on the Pine Ridge Indian Reservation, nestled between the Badlands of South Dakota and the Nebraska border, and the weather was unseasonably pleasant. My colleague, who had followed office procedure in substituting her car for a rental to travel the poorly maintained roads into a high-crime area, informed me that we were in the second-poorest county in the United States. Most Americans would probably be shocked to encounter a territory larger than some states in the heart of their country where living standards are comparable to developing nations. I stopped checking to see whether my cell phone service had returned when I learned that many residents had no running water.

My two colleagues and I paid visits to a variety of offices supported by federal funding, from veterans' posts to schools and job training facilities, arriving in the early afternoon at a community medical center. While I was preparing to leave the men's restroom there, a local man came in, went straight to the sink, and determinedly scrubbed his arms up to his elbows in a thick lather with hot water. I assumed that he was a doctor prepping for surgery, and I felt sorry that he had to wash in a sink in the lobby toilet area.

"We just got mail from Washington, D.C.," he explained after giving me a cordial hello. "You can't be too careful."

It was Monday, November 5, 2001. Even in this remote patch of what seemed like a Third World country, they were afraid of mail from D.C. I didn't have the heart to tell him why I was there.

RECOGNIZED THREATS AND MISSED OPPORTUNITIES

Exactly three weeks earlier, I had gone to work as usual in my cubicle in suite 509 of the Hart Senate Office Building, a block away from the US Capitol. It was the office of the senior senator from South Dakota, Tom Daschle, where I had been working for a year and a half during a particularly turbulent period of American history. The disputed 2000 presidential election had resulted in Daschle's sudden elevation in public prominence as Senate minority leader, a development reinforced a few months later when another senator's party switch made him the majority leader. Many conservative South Dakotans were increasingly outraged by his actions as de facto leader of the Democratic Party, but these developments brought attention from elsewhere as well: One day a bakery that was unfamiliar to me sent our office a box of cookies as congratulations to Tom Daschle on becoming majority leader. When I suggested that perhaps they should be checked in case the baker was not actually a political supporter, my colleagues laughed as they ate the delivery. It was still summer 2001.

There had been little laughter around Capitol Hill in the month following September 11. The memory of two evacuations (one occurred two days after the attacks because of the discovery of a suspicious package), the smoke across the river billowing from the Pentagon, and the televised horror in Manhattan were far too fresh. The one hijacked plane that had not reached its target had been intended for the Capitol, and it had been too close. Although the flags that had been flying at half-staff for weeks had finally just been raised again, there were still no flights allowed into National Airport, and we were preparing for war in Afghanistan. It was evident that the old normalcy would not be returning.

And something else unsettling had begun to happen. First in a tabloid newspaper office in Florida, and then in national television network offices in New York City, individuals were sickening and even dying from anthrax, which media coverage described as a rural disease that normally afflicted farm animals. This was apparently yet another terrorist attack and, with evidence mounting that the deadly *Bacillus anthracis* spores had been delivered via hate mail, concern among Daschle's staff turned to the possibility of an anthrax letter arriving in our office. Despite the senator's high profile and the unprecedented volume of (mostly angry) mail we were receiving, Chief of

Staff Pete Rouse told everyone to relax: "I seriously doubt that anyone is targeting the Hart Building," he said on Friday, October 12.

My first hint that something unusual was happening on the morning of Monday the 15th was when I realized that my neighbor in the cubicle across from mine had never returned from going to check her mailbox. I had earlier gone up the internal office staircase to the mail room on the sixth floor of the building, and I was sorting through letters and e-mails at my desk when I realized that she had been gone for forty-five minutes.

Shortly after this, the office managers began to inform everyone, individually and quietly, that there had been an incident in the mail room, with a letter and white powder similar to the one that had been sent to NBC News. It was probably a hoax, but we had to take precautions. The Capitol police had already been on the scene, and we soon learned that we were quarantined until further notice. The water cooler was empty that morning, which became a source of discomfort in the subsequent hours after the air conditioning was shut off, but at least the toilet wasn't broken like it had been a couple of weeks before.

The news soon began to filter out. President George W. Bush, politically our archenemy but at the moment the historically popular commander in chief, announced at a televised press conference that we had received real anthrax, something that had not been confirmed to us in the office yet. E-mails came in from friends asking me if I was all right; my grandparents called. I had been instructed to give out no information, so I simply said that I was OK and would let them know more. My fiancée called, hysterically demanding that I get out of the office immediately. I couldn't simply run out on my colleagues, I informed her, and besides, "There are men blocking the door with machine guns requesting that we not leave."

There was nothing to do but sit and wait. I thought back to a paper I had written three years earlier for a professor named Andrew Krepinevich, whose policy and academic work centered on gaming future threat scenarios. When he told me that he had never seen a really good analysis of the military implications of biotechnology, away I went. One observation was,

> Due to the ready availability of biotechnology information, including instructions for creating anthrax weapons available on the Internet, a [domestic] terrorist attack is perhaps likelier to occur within the next several years. . . . Among biological agents used for warfare, anthrax is the gold standard. . . . It is cheap, stable, can be stored almost indefinitely as a dry powder, and has an incubation period of nearly a week, meaning that detection is not likely until long after distribution. Preventative vaccines do exist, and recovery through antibiotics is possible, although often not after symptoms manifest themselves. For these reasons, anthrax is currently the single greatest biological threat facing the United States.

And yet, I was not particularly frightened while being far closer to the subject than I had ever anticipated. Antibiotics had proven effective on people in the NBC News office even after they had developed symptoms, and because it was reasonable to assume that this was the same strain of anthrax, and it had been caught at the very beginning, I saw no reason for alarm.

Clearly the authorities were not alarmed either: a pizza delivery man had been sneaked in via a back staircase to bring us lunch (he later had to be tracked down, his steps retraced, and given a supply of ciprofloxacin), and we were sent home that afternoon without any antibiotics. We were told that anyone on the fifth floor of the office would not have been exposed, and the Capitol physician's office therefore initially insisted that we did not require testing for spores. We were told to leave everything in the office, but I figured that we might not be back and took my briefcase and important personal effects with me. Many of us insisted on testing and, after sneaking out the back staircase used by the pizza man, we went to the physician and received nasal swabs with six-inch-long q-tips. I could feel pressure from the stick behind my eyeball and compared the experience to an alien probe.

I drove home to where my fiancée was waiting, and it didn't occur to me until hours later that perhaps I shouldn't have embraced her: Who knew what was on my suit? It went in a bag, and the seat cushion it had been lying on went in the washing machine, which spilled all over the kitchen floor because it had been overfilled.

I received two calls from supervisors that night to see if I was all right, to make sure that I wasn't talking to the press, and to tell me to come in the next day for a precautionary three-day supply of Cipro. I reentered the Hart Building the next day and stood in line with hundreds of staff waiting for their three pills, most of whom were probably quite chagrined when the entire building was shut down as a health hazard shortly thereafter. Except for those who experienced psychosomatic symptoms, for them that was the end of the ordeal. For more than twenty others who came into contact with mail contaminated by the Daschle envelope, it was about to begin.

And for my colleagues and me, it was far from over. The next day we had a long meeting in the Capitol building itself and learned that a number of the staff on the fifth floor had tested positive for anthrax exposure, and more positives were identified as the day went along. I noticed that, unusually—and particularly so during that time of heightened security—there was someone circling near the Capitol in a hang glider. The news coverage that morning had focused on the threat of a radiological "dirty bomb," and I kept waiting for the glider to fly right into us, but eventually it disappeared. We were all sent home with two-month supplies of Cipro, but I was already experiencing the upset stomach that we had been warned could be a side effect. A week later, when I was told that I was being sent to our Rapid City,

South Dakota, office as a result of lack of work space at the Capitol, my symptoms were getting steadily worse.

All of the staff was on antibiotics for 103 days, although many of us had to conclude our course on something other than Cipro, which had never been previously given to humans for a duration of months. We also eventually received three rounds of the anthrax vaccine used by the military, which had only been tested prophylactically up to that point but was recommended because of uncertainty about the antibiotics.

Ultimately, the greatest thing that many of us had to fear was Cipro itself. One of my colleagues had an appendicitis that had been scheduled for surgery but actually reversed itself in the weeks under the drug. A staffer in an adjoining office, who was put on Cipro as a precaution, "now walks with a cane because the powerful antibiotic nearly destroyed his Achilles' tendons. After he sued the drugmaker, Cipro now comes with a warning about its potential effects on tendons" (Pierce, 2011).

The Daschle staff had been warned in an early meeting that Cipro could cause an Achilles' tendon rupture, and mine started hurting almost immediately thereafter, so I assumed it was psychosomatic. Soon, however, I was finding it increasingly difficult to walk. But I was more preoccupied with my upset stomach, which had taken a turn for the worse: I was now filling toilets with significant volumes of blood a couple of times a day. That is what I had been doing in the men's room in Pine Ridge when the man terrorized by the possibility of cross-contaminated mail came in to scrub. After forty days on Cipro, I was allowed to switch to something else, and the symptoms soon subsided, although the bleeding continued intermittently for years.

Initially I had been told by the physician attending us that Cipro did not cause internal bleeding, and he insisted that I see a specialist. This doctor examined me for fifteen seconds, stated that his opinion was that the Cipro could indeed be responsible, and then sent me a bill for $240. I refused to pay out of pocket because I had only seen him at the insistence of the doctor overseeing our care. So I was told to go to the Senate Employment Office and file a claim with the US Department of Labor for compensation for a work-related injury. My claim was rejected, but when I complained to our office manager, she took all of my paperwork and said it would be taken care of. After that I stopped receiving overdue payment notices from the specialist. As far as I know, I was the first person in history to receive workers' comp benefits for a bioterrorist attack.

Ultimately the Federal Bureau of Investigation identified Dr. Bruce Ivins, the lead anthrax defense researcher for the US Army, as the sole perpetrator of the 2001 attacks (termed "Amerithrax" by the FBI because of the domestic source of the spores). It was subsequently disclosed that Ivins had a known history of severe psychological disturbance but was still permitted unrestricted access to deadly pathogens. Except for one woman whom he had previously threatened, none of Ivins's colleagues in the biodefense establishment had named him as a potential suspect during the FBI's seven-year investigation (US Department of Justice, 2010). Either it did not occur to these scientists and security officials that the man who made death threats to colleagues might be the source of the spores that came from their lab, or they chose not to implicate him and, perhaps by extension, their professional enterprise.

BIOTECHNOLOGY AND THE TWENTY-FIRST-CENTURY INTERNATIONAL ORDER

As troubling as the biological attacks of the fall of 2001 were for the victims and their families, they also raised a host of security concerns that have persisted long after the last round of anthrax treatments were administered. Fundamentally, they confirmed that various fields of biotechnology have serious ramifications for national, international, and human security. Was this an isolated occurrence or part of a broader trend toward the use of biological armaments? What, if any, sort of protective measures should be taken on behalf of the public? Requiring every citizen to receive vaccinations or ingest medications such as Cipro? Limiting access to previously open academic research into diseases, or even the human immune system?

What would an appropriate response have been if the source of the spores had been identified as a foreign government—a nuclear strike to deter other adversaries? An in-kind biological attack? Would bioweapons be legitimate to use in the new War on Terror? Would it be more or less legitimate to employ agents that attacked agriculture—such as the opium-producing poppies that the Afghan Taliban used to finance their activities—than humans? How about weapons that did not permanently damage people in target areas but incapacitated them by disrupting their basic neurological functions? On the flip side, could such technologies be used to safeguard American troops on the battlefield and ensure military superiority?

The terrorist attacks of 2001 raised public consciousness about the potentialities of these lines of research. They also spurred major investments by governments, militaries, and the private sector into biodefense research, as well as in other, different lines of biotechnology for national security purposes. And they raised the responsibility to ask questions about what doctrines are being developed for the use of these biotechnologies—before they are deployed on the battlefield or turn up in unexpected places, like the anthrax spores sent to the US Capitol that had been created in a US Army biodefense lab.

Introduction

Biotechnology beyond Germ Warfare

The biotechnology revolution will have implications for security that will probably exceed those of the nuclear and information revolutions that preceded it.
—Ashton Carter, US Secretary of Defense

Our responsibility is to develop the technology. How it is ultimately used will be determined by the military. . . . Our singular mission is the creation and prevention of strategic surprise. It would be inconceivable for us to do that work if we didn't make people excited and uncomfortable at the same time with the things we do. . . . We can't simply close our eyes and pretend that the technology isn't advancing. It's advancing.
—Regina E. Dugan, Director, Defense Advanced Research Projects Agency (DARPA)

Human beings have used biotechnology in one form or another in warfare since the dawn of civilization. And yet bioweapons remain the least familiar category of arms to most students of international relations. As opposed to a technology like air power that has been used strategically by major powers in modern times, it is difficult to gauge the impact of bioweapons. But with a wide range of biotechnologies from soldier enhancement to genetic weapons in development or already deployed, new biotechnologies are likely to significantly transform military power in the twenty-first century. It is therefore worth examining how the full scope of biotechnology impacts international security, from the rivalry of great powers to the human security concerns of individual citizens.

1

This book is intended to present a thorough but accessible introduction to bioweapons and biosecurity challenges and the ethical and policy dilemmas that come with them. It is written primarily for readers with an interest in international relations and strategic studies who do not necessarily have a technical background or an in-depth familiarity with biological weapons programs. Hopefully it will also engage readers with a medical or engineering background who are interested in how security decision makers are receiving discoveries that may not have been imagined with military purposes in mind.

It therefore introduces, in the first two chapters, a broad overview of the history of biological weapons and advances in genetic engineering for audiences who are likely not familiar with these developments, before addressing the scope of current research and the normative questions surrounding it. Just as international security texts have not needed to detail the physics of nuclear fission or the effects of radiation poisoning to analyze the import of nuclear weapons, I also avoid describing the physiological effects of the weapons or how unpleasant the pathogens look under a microscope.

This book challenges the conventional wisdom that biotechnology represents a security threat to major powers because it allows weaker actors to attack them at lower cost. I argue instead that the most significant developments underway in biotechnology will actually retrench the hegemons of the international system with advanced weapons. Although the dissemination of technology and lower costs for buying it commercially will permit other actors to catch up with particular technologies, they will already be behind newer developments.

It also contests the assumption that the security impact of biotechnology can only be examined speculatively or as a counterfactual of strategies that were not employed. Not all of the emerging technologies described will become factors in twenty-first-century international security, and some are likely only intended as aspirations by their R&D teams. But others are on the way or have already been used on the battlefield. The future is not only arriving now; it has been here for some time already (Singer, 2010: 300).

THE SCOPE OF BIOTECHNOLOGY

Biotechnology has existed throughout recorded human history, dating back to the fermentation of wine and bread leavened with yeast, but the term itself was not coined until 1919 by Hungarian engineer Karoly Ereky (Melson, 2003: 2). The US Office of Technology Assessment defines *biotechnology* as "any technique that uses a living organism, or parts of organisms, to make or modify products, to improve plants or animals, or to develop microorganisms for specific uses." The European Union employs the definition of the international Organization for Cooperation and Economic Development: "The application of science and technology to living organisms, as well as parts, products and models thereof, to alter living or non-living materials for the production of knowledge, goods and services" (European Commission, 2013). And the United Nations describes biotechnology as "any technological application that uses biological systems, living organisms or derivatives thereof, to make or modify products or processes for specific use" (United Nations, 1992). Other definitions include material patterned after living organisms but not necessarily using them as components (biomimetics).

The types of biotechnologies that influence international security are therefore far broader than attacks using pathogenic diseases as weapons. Even before the agricultural revolution, humans used animal bones as tools and weapons to provide them with an asymmetric advantage over competitors. This strategy has continued unabated, but now these tools of defense and conquest also use genetic engineering and materials sciences to impart the desired properties of animals and plants directly into manufactured materials.

The firms and research labs that create them have experienced rapid growth in recent years, with the American biotech sector expanding from an $8 billion industry in 1992 to one that was estimated at over $39 billion in value a decade later and that was spending nearly $18 billion annually on research and development. By 2004, there were over 1,500 biotech companies in the United States alone, of which only 20 percent were publicly held (Shelton, 2005: 5–6). By 2013, nearly 1.5 million Americans worked in the biotechnology sector, with more than 5 million others employed in related industries (SelectUSA, 2013). China's biotech sector is reported to be worth over $600 billion (All China Biotech Conference, 2014).

Various biotechnologies also have both direct and indirect international security implications. The National Research Council's Committee on Advances in Technology and the Prevention of Their Application to Next Generation Biowarfare Threats (hereafter Committee, 2006) noted that biowarfare and biodefense were, prior to the twenty-first century, concerned with naturally occurring, and later genetically engineered, pathogens and toxins used in biological warfare. However, "other fields not traditionally viewed as biotechnologies—such as materials science, information technology, and nanotechnology—are becoming integrated and synergistic with traditional biotechnologies in extraordinary ways, enabling the development of previously unimaginable technological applications" with military dimensions (Committee, 2006: 1).

A BROADER VIEW OF BIOTECH

Still, while academics researching biological sciences and defense have continued to produce significant and advancing work during the twenty-first century, the international relations discipline has remained focused almost exclusively on traditional modes of biological "germ" warfare through the history of state bioweapons and biodefense apparatuses rather than on the implications of wider applications of biotechnology. The literature can be divided into subfields. One strand focuses on the history and technical aspects of the potential threat of bioterrorism by non-state actors. Another category of works has likewise presented highly descriptive accounts of state efforts related to bioweapons in production, defense, and arms control efforts. More recently, researchers have used anthropological and organizational behavior approaches to study the conceptual frames of bioweapons scientists and policy makers, explaining their operational and doctrinal constraints in an approach reminiscent of Graham Allison's (1971) study of imperfect information and national security decision making.

One common feature of the bioweapons literature is that many, if not most, of the authors have backgrounds in biological sciences and have worked in governmental or intergovernmental agencies responsible for biodefense or counterproliferation. The bioweapons literature is therefore far more technical than perhaps any other subfield of international security, but it remains undertheorized. It is distinctive from, for example, the foundational neorealist literature of the Cold War era on nuclear weapons, written by scholars with no prior background in nuclear engineering or policy making. The theoretical underpinnings of this book are largely adopted from constructivist works on chemical warfare, nonlethal weapons, and the ethics of warfare.

While there is a substantial and growing literature on legacy twentieth-century bioweapons programs and on the potential for bioterrorism, little attention has yet been paid to other advanced biotechnologies or their intentional development for warfare. For the most part, the literature on how developments in other life sciences might affect security is concerned with the problem of dual-use technologies. The assumption is that scientific research for human health and peaceful commercial purposes might be perverted for applications as weapons, with World War I germ warfare programs based on the work of Louis Pasteur cited as an example (Tucker, in Danzig and Tucker, 2012: 1–12).

Arguably, applications of neuroscience have received the most attention as potential disruptive battlefield technologies, particularly programs that are intended to speed up the reflexes and cognitive processing of soldiers and pilots through electrical or chemical stimulation (see Dando, in Pearson, Chevrier, and Wheelis, 2007; Huang and Kosal, 2008; and Moreno, 2006).

But analysis of the potential harmful effects of these technologies occurs in the context of their subsequent adoption by adversaries such as rogue states or terrorists and not by their developers. For example, Moreno (in Danzig and Tucker, 2012: 228) discusses evidence that transcranial magnetic stimulation can disrupt brain functions governing moral judgment. He notes the potential that this technology "could cause subjects to suspend their moral judgment and behave according to some grossly utilitarian calculus, such as 'If you carry out a suicide attack, you will send a message to our foes and save the lives of many others.'" But such treatments could also be used by regular military forces to make soldiers more effective warfighters. As ethicists who debate human enhancement (discussed in chapter 2) argue, the moral calculus about the use of such technologies flips quickly when they are framed as saving the lives of your own people—and particularly when the competition is doing it too. What is the appropriate doctrine then?

While this book examines potential threats from dual-use biotechnologies, it also seeks to analyze the impact on international security of dedicated-use technologies that have been and are currently being developed expressly by and for military purposes. As with other at-the-time advanced technologies, such as jet aircraft and radar, they are likely to subsequently diffuse to commercial markets for public use and misuse. But in those cases, the militaries of the major powers have retained an unmatched competitive advantage because they retained the greatest concentration of resources and investment in R&D, and the same should hold true for biotechnologies as well.

ARGUMENT AND APPROACH

What some are describing as an unfolding "biotech revolution" does not appear to be a revolution in the traditional sense of radically rearranging distributions of power. In the mid-twentieth century, the agricultural Green Revolution increased the long-term power of some Third World states by enabling them to grow crops more efficiently and enlarge their populations considerably. Instead, in the twenty-first century, the states that are best positioned to reap the benefits of new biotechnologies are those that are already scientifically advanced, rich, and powerful. Many security analysts have stressed the advantages conferred on less powerful actors by biotech, particularly the possibility of devastating attacks by rogue states once they gain the capacity to produce their own biological weapons. On balance, however, it is advanced industrial states with strong military research programs and thriving commercial sectors that will emerge in stronger positions because they will be able to implement advances that rogue and non-state actors cannot begin to match. And they are beginning to do so already.

Biowarfare and bioterrorism are particular applications of biotechnology that involve the production and deployment of harmful biological substances for purposes of political violence. However, the important security dimensions of biotechnology are broader than the germ warfare programs created by major powers in the twentieth century, or the threat of such attacks by non-state actors in the twenty-first. As noted, the biotech field actually represents a collection of sciences, including genetic engineering, enzymology, fermentation, and other attempts to enhance the performance of organic material, either chemically or genetically. Over the course of the coming quarter century, these developments will have profound effects on human health, food production, and synthetic material development. The implications for military applications will ultimately be as profound. In particular, genetic engineering offers the likelihood of not only providing effective counters against bioweapons used by rogue states and terrorists, but for attacking the very biochemistry of hostile enemies. Programs to develop such advanced weaponry are already yielding results, raising a host of questions for human security and the ethics of warfare.

Biotechnology can alter the workings of the genetic codes of the cells that comprise all living organisms (Purdue University, 2001a). Rather than just anthrax attacks and defenses, various militaries are already integrating diverse biotechnologies into their force planning that will enable them to monitor the well-being of soldiers in the field, improve their memory retention and reaction times, rapidly heal serious injuries, and keep food fresher for longer deployments—not to mention the ability to imitate the physiology of geckos, allowing soldiers to scale walls like comic book superheroes (all of which are Department of Defense–funded programs presented in chapter 3). Likewise, the promise of fuel production from bacteria and the eradication of disease through genetic therapy portends a new era of plenty—in the societies that have the resources to develop it. What do these coming developments in power projection capability for hegemonic actors portend for the future of the international system?

To explore these challenges, I begin in chapter 1 with a broad overview of the uses of biotechnology in armed conflict since the beginnings of recorded history, culminating in the massive state biological weapons programs of the twentieth century and the threat of the proliferation of loose bugs. Chapter 2 examines the advent of genetic engineering and what it has meant for both the power and promise of biotech, from potentially unstoppable bioweapons to potentially limitless electrical and computing power. I examine the developments in both chapters from the perspective of the international relations discipline rather than life sciences.

Chapter 3 describes how emerging developments in biotech are already reshaping conventional military forces and the potential they offer for asymmetric warfare, as well as commercial-sector advances and how these affect the distribution of international power. Finally, the last two chapters address questions raised by these advances: chapter 4 addresses developments in homeland security, including leakage and proliferation originating with state biodefense programs or academic and corporate research, and the equitable distribution of biodefenses within diverse societies. Chapter 5 examines the role of biotechnology in both maintaining and threatening norms of just warfare, such as the role of nonlethal weapons, and bioethics, including the privacy of health information.

The field of biotechnology is poised to make rapid and significant imprints on homeland, national, and international security policies, and it is vital to understand its existing history and the peril and promise of its future. From the billions of dollars already spent by the United States to develop augmented super soldiers to the potential effects of emerging technologies on the global energy and agricultural markets, decisions are being made that will affect the lives of individuals even further from the centers of political and commercial power than a Pine Ridge medical clinic worker terrorized by anthrax in Washington, D.C. Although the net effect will be to re-entrench the most powerful actors in the international system of the twenty-first century, biotechnology will subtly but inescapably reshape international security in ways that affect the entire human race.

Chapter One

A History of Biotechnology in Warfare

In *The Tempest*, the tale of a man who seeks to build order using a unique knowledge of nature given to him by adept specialists, William Shakespeare wrote, "What's past is prologue." This chapter offers an extended prologue to the development of novel advanced military biotechnologies. It examines how actors throughout international history, from great powers to rogue states, progressed from sporadic attempts to use naturally occurring diseases and toxins, and the far more devastating inadvertent pandemics caused by their military and economic hegemony, to the industrial-scale production of bioweapons in the twentieth century. The result of this legacy is the life sciences research infrastructure that is preparing the biotechnologies of the twenty-first century for military purposes.

THE THREAT OF BIOWEAPONS

The term *biological weapon* (shortened to *bioweapon* or *BW*) generally refers both to living organisms and to the nonliving biochemicals and poisons that they secrete. The former category includes viruses, which are not technically living organisms because they cannot reproduce on their own but that have the ability to replicate by hijacking the cells of their hosts. Some scholars argue that biological agents could also include more complex organisms, including biologically enhanced soldiers, or at least engineered substances and technologies within their bodies that make them more lethal to others (Dvorsky, 2013).

Substances in the poison category are typically grouped together under the label of toxins and include a range of deadly substances, from snake venom to ricin to botulinum toxin. These biological weapons differ from chemical weapons in their preparation and their greater lethality (Maurer, in

Maurer, 2009: 106–107).[1] For example, *Clostridium botulinum* bacteria can be grown and harvested in laboratories to produce botulinum toxin, a natural substance ten thousand times more lethal by quantity than deadly chemical weapons such as the VX nerve agent (Koblentz, 2009: 5, 64).[2]

The World Health Organization defines biological agents "as those that depend for their effects on multiplication within the target organism and are intended for use in war to cause disease or death in man, animals or plants; they may be transmissible or nontransmissible." However, the international Biological Weapons Convention, enacted in 1975, does not define what are considered to be biological agents or toxins (Roberts, 2003: 5). Additionally, subsequent innovations in genetic engineering offer novel approaches through the disruption of the human body's basic functions by rewriting the target's genetic code. Although these new "direct effect" weapons would be delivered by vectors (engineered synthetic viruses), they would effectively turn the body's own regulatory systems against itself rather than using foreign substances.

Utgoff (in Roberts, 1993: 28–30) describes the functional requirements of an effective biological weapon as (1) deadly or debilitating, (2) relatively fast acting, (3) highly contagious if a pathogen, (4) predictable in its effects, (5) able to persist in air or water long enough to cause widespread epidemics, (6) not susceptible to common treatments for infection, (7) not readily destroyed through purification methods, and (8) not easily susceptible to antidotes or prophylactics.[3] Historically, however, given the natural attributes of the infectious agents, few weapons could be produced that met the desired qualities, and in sufficient quantity to be useful.

From a strategic perspective, bioweapons are described by scholars as favoring offense-based strategies. They can be deployed anonymously and invisibly, disguising the identity of the attacker, and even the fact that an attack has occurred until after the victims have already been debilitated. But because they are invisible and their effects are not instantaneous, they may also fail to provide deterrence (Koblentz, 2009: 21). For this reason, rogue states and violent non-state actors such as terrorist groups may be tempted to strike with them, while the major states that would be their targets cannot convincingly threaten in-kind retaliation as they did during the Cold War nuclear standoff.

In 1996, the Pentagon Office of Counterproliferation and Chemical and Biological Defense argued that

> Biological weapons are the most problematic of the Weapons of Mass Destruction. They have the greatest potential for damage of any weapon. They are accessible to all countries, with few barriers to developing them with a modest level of effort. The current level of BW sophistication is comparatively low, but there is enormous potential—based on advances in modern molecular

biology, fermentation and drug delivery technology—for making sophisticated weapons. (Preston, 2009: 8, 182)

Scholars who argue that BW will be a significant source of international instability in coming decades focus on the probability that terrorists, states that sponsor terror, cults, or psychotic lone actors will be increasingly able to acquire or manufacture them. These concerns essentially divide into "supply" and "demand" camps based on the authors' perception of the value of defensive military biotech research.

On the supply side, those who contend that governments are spending on unnecessary and hazardous weapons programs in the name of homeland security argue that the expansion of such research is inevitably creating more opportunities for the accidental release of pathogens and for stockpiles of poorly secured materials to be misappropriated. In this view, the purveyors of biotech solutions to national security are actually sowing the seeds of their own destruction.

Alternatively, advocates of these programs point to the demonstrated willingness of violent non-state actors to attempt to obtain and use WMD as evidence of a growing demand for bioweapons. They also note that the proliferation of genetic engineering technology to the commercial sector has now made it possible for individuals to produce deadly organisms with biological material and processing equipment that can be obtained by mail order for the cost of a few thousand dollars. Additionally, as biotechnology and technical knowledge become more widespread, they fear that pathogens will be manipulated or mutated into particularly virulent forms against which there are no ready defenses available. Under the most extreme scenarios, the survival of the entire human race hangs in the balance, not to mention global political and economic order.

With this capacity for silent destruction, biological threats are frequently identified as "an increasingly serious and complex threat to national security. [A] recent National Intelligence Estimate identified the threat of bioterrorism as the intelligence community's most significant WMD-related concern" (Gronvall et al., 2009). Thomas Preston (2009: 182) argues that biological weapons pose a far greater security threat than even nuclear arms because they are more easily available and more difficult to control using tested counterproliferation strategies, since so many components have legitimate commercial uses as well. Furthermore, Gregory Koblentz (2009: 19–21) notes that "with genetic engineering, traditional BW agents can be made potentially deadlier, more resistant to antibiotics and vaccines, and better able to avoid detection systems. In addition, harmless microorganisms can be transformed into deadly ones with novel properties."

Further, not all research laboratories practice strict security measures, with some allowing individuals who have not had background screenings access to poorly guarded samples of dangerous specimens. Even in the most closely guarded laboratories, the human element is a factor, as in the case of the Amerithrax attacks. Still, the research sector for defense applications of biotechnology continues to build on the foundations of military bioweapons programs. As Amy Smithson (2011: 6) postulates, as chemistry begat the hallmark weapons of World War I, and physics the most notable weapon of World War II, "the militarization of the fruits of modern biology may be the distinguishing feature of future conflict."[4]

EARLY BIOWARFARE

Throughout recorded history human beings have used a variety of living elements or organic material to augment the destructive power of their weaponry. Although the germ theory of disease transmission was unknown in the Western world until the 1700s, it had nonetheless been long recognized that some diseases were transmissible, and infected items were used by armies both on the battlefield and against civilian populations. At other points large organisms, including live animals, were employed directly as weapons. And some armies, usually unknowingly, carried diseases with them that would decimate entire populations more efficiently than they could ever dare to dream by using force of arms, reshaping the international order for centuries afterward.

Antiquity and Beyond

Potentially, most biological attacks throughout history have gone unrecorded because the victims were unaware that they had occurred or were unable to report them (Wheelis, in Geissler and Moon, 1999: 9). Despite this, there is no shortage of documented attempts to employ bioweapons in the ancient and medieval worlds. As far back as the seventh century BCE, the Assyrians poisoned the wells of their enemies with an ergot fungus from rye grains that produced hallucinations, seizures, and death. That those affected seemed to go mad further served as a form of bioterrorism perpetrated against the remaining enemies (Texas Department of State Health Services, 2007).

Records of the Ancient Greeks polluting their enemy's wells and drinking water supplies with animal corpses date to 300 BCE, and later the Romans and Persians would adopt the same strategy to harness botulants during sieges (Roberts, 2003: 15). Given that germs were unknown at the time, the ultimate objective in these instances would have been simply to deny the enemy potable water rather than to transmit infections (Wheelis, in Geissler and Moon, 1999: 9). During the siege of Cirrha in 590 BCE, Athens reportedly contaminated that city's aqueduct with purgatives from hellebore roots to incapacitate the defenders (Texas Department of State Health Services,

2007). "At Themiscrya, another stubborn Greek outpost, Romans tunneling beneath the city contended with not only a charge of wild beasts but also a barrage of hives swarming with bees—a rather direct approach to biological warfare" (Tharoor, 2009).

The Romans were hardly averse to incorporating biotechnology into their war machine. The catapulting of bee and hornet nests was apparently so widespread that it contributed to a shortage of hives throughout the empire. Legionnaires responded to guerilla attacks in Asia Minor by poisoning wells in the areas where the insurgents operated. And poisons were a vital element of state security: "A famous serial poisoner named Locusta was employed by Emperor Nero to organize a school of poisoning where she could tutor others and conduct experiments aimed at determining how to poison and how to defend the person of the Emperor against poison" (CBWInfo, 2005).

While setting a precedent for the state BW programs of the twentieth century, the Romans themselves were victims of biowarfare efforts on numerous occasions. Among them, the famous general Pompey suffered the loss of many legionnaires on an expedition to Pontus (modern Turkey) when the local inhabitants fed them with toxin-laced honey produced from local flowers, leaving them suffering from nausea and hallucinations and open to a counterattack.[5] And during the Battle of Eurymedon in 190 BCE, Hannibal won a naval victory over King Eumenes II of Pergamum by catapulting pottery jars containing poisonous snakes onto his enemy's ships, causing them to retreat (Ackerman and Asal, in Clunan, Lavoy, and Martin, 2008: 187; CBWInfo, 2005).

Chinese sources dating to 400 BCE describe how smoke from burning mustard plants and wolf excrement was funneled into tunnels to discourage the besieging army (CBWInfo, 2005). Other sources describe Chinese countersiege tactics involving the use of boiling human waste from toilets (Wheelis, in Geissler and Moon, 1999: 11).

Forms of biological warfare continued across Europe through the Dark and Middle Ages. Viking *berserkergang* warriors may have used psychoactive agents to attain fits of uncontrollable rage in which they demonstrated prodigious strength and indifference to wounds. Invading Vikings were killed by Scots who offered them barrels of beer laced with the belladonna plant, which was also used as a common tool of assassination in the political intrigues of the Italian city-states (Bevan-Jones, 2009). At the battle of Tortona, Italy, in 1155, the Holy Roman Emperor Fredrick Barbarossa used human corpses to poison his opponents' water supply. In 1485, clashes over royal succession in the Kingdom of Naples reportedly led the Aragonese faction to send their Anjou opponents wine laced with the blood of leprosy patients. Later, Napoleon would attempt to start an outbreak of swamp fever in Mantua (Roberts, 2003: 37).

Perhaps the most dramatic and colorful employment of biowarfare during this period was the tactic of catapulting infected corpses over city fortifications during sieges. The earliest reliably documented incident occurred during the Hundred Years' War in the siege of Thun l'Eveque in 1340. One eyewitness reported that siege engines were used to catapult decomposing animal corpses into the city in the summer heat, causing the defenders to withdraw, although no outbreaks of disease among the besieged were reported in this instance (Wheelis, in Geissler and Moon, 1999: 10–11).

Among these tales, one that appears frequently in the literature on bioweapons is the allegation that the Tatars (Mongols) attacked Caffa (present-day Feodosia on Ukraine's Crimean Peninsula) in 1347, catapulting their own soldiers who had expired from plague (*Yersinia pestis*) over the city walls (Gerstein, 2009: 45, 161). Despite the repetition of this story, there is no evidence that these events actually transpired in this manner. The account originated from a single report that appeared in the papers of a lawyer who worked in a town near Genoa who claimed to have heard the story from Sicilian sailors (Weatherford, 2004: 244).

Bacterial Colonies

The outbreak of the particularly virulent plague that killed as much as half of the Eurasian population in the fourteenth century was indeed transmitted by the Mongol Empire, but in a fashion more terrifying and meaningful for the globally interconnected world of the twenty-first century than catapulted cadavers. Instead, it was precipitated through the economic and political system erected by the Mongols that stretched from the Pacific to the Mediterranean.

Genetic evidence indicates that the plague began in southern China. Ninety percent of the population of Hopei Province died in 1331, with the population of China falling by between half and two-thirds within twenty years. The following year, the ruling Mongol imperial family collapsed under the strain of the plague, ultimately with four rulers in as many years, including a seven-year-old who held the throne for only two months. The facts that the fleas that carry the plague do not normally attack humans, that the smell of the nomadic warriors' horses repelled them, and that the Central Asian plateau had a low population density meant that most Mongols had little to fear.[6] But the fleas thrived in sacks of grain and clothing carried not only by the warriors, but by merchants like Marco Polo who operated under the common markets made possible by Mongol governance. As Jack Weatherford (2004) notes,

> [China was] the manufacturing center of the Mongol World System, and as the goods poured out of China, the disease followed, seemingly spreading in all

directions at once. . . . The plague was an epidemic of commerce. The same Mongol roads and caravans that knitted together the Eurasian world of the thirteenth and fourteenth centuries moved more than mere silk and spices. The roads and way stations set up by the Mongols for merchants also served as the inadvertent transfer points for the fleas and, thereby, for the disease itself.

The plague caught up with Yanibeg Khan when he laid siege to Caffa, which was a Genoan merchant port that served primarily as a trafficking point for Russian slaves being sent to Egypt. Although Yanibeg lifted the siege when plague broke out in his ranks, the disease spread from his camps to the port. The fleeing Genovese carried the plague to Constantinople, from which it spread to Africa and Europe, and to the edges of the known world, ultimately destroying the Norse population of Greenland.[7]

The consequences of an outbreak that became a pandemic through globalization provide a salutary lesson. Seven hundred years after Europeans erupted in lesions called buboes, the word is still taught to Western children to refer to injuries. The continent lost nearly half of its population in the pandemic, as compared to 19 percent civilian casualty rates in Poland and Ukraine, the worst affected countries during World War II (Weatherford, 2004: 241–245).

The Europeans in their turn exposed the populations of the New World to pathogens against which they had no acquired immunity, reducing their ranks by estimates of 90 percent and enabling their conquest by small bands of soldiers and fortune seekers. The Spanish conquistadores who caused the downfall of the Aztec and Inca Empires and numerous tribes of the Americas were largely unaware that they carried smallpox and other pathogens, and so, like the Mongols, they were not consciously practicing biowarfare (Texas Department of State Health Services, 2007).[8]

However, relatively soon after, biotech developments in disease prevention inspired another colonial power to turn the deadliest pathogen known at the time against restive natives: smallpox. Beginning with a scientific journal article published in 1714 and continuing with the activism of survivor Lady Mary Wortley Montague, efforts began in Britain to contain the disease by variolation, or immunization, of previously unafflicted patients with scabs from smallpox victims. Montague had seen the practice, by which the presence of weakened or dead smallpox specimens stir the creation of antibodies that will block the patient from being susceptible to the infection, practiced while her husband served as ambassador to the Sublime Porte of the Ottoman Empire. Just as the Black Plague had spread to the West through Constantinople, the means of eradicating smallpox now emanated from Istanbul (CBWInfo, 2005).

In 1770, when Edward Jenner introduced the biotechnology of vaccination, building immunity through exposure to another, milder pathogen to stimulate antibody production—in this case cowpox samples—"the Army seized upon it almost immediately. All troops were ordered to be vaccinated. This measure prevented soldiers from developing smallpox" and prevented them from spreading it through movements through the countryside (Marble, 2010).

However, as with defensive research that produced the 2001 anthrax case and the potential for other threats in the twenty-first century, scientific knowledge intended to prevent disease in the military was soon used as a weapon. Once it was protected by vaccination, the British Army showed no compunction about spreading smallpox against opponents. One of the best-documented examples occurred in the North American theater of the Seven Years' War between the great powers of Britain and France, remembered in the United States as the French and Indian War. In 1763, the final year of the conflict, a cross-tribal uprising remembered as Pontiac's Rebellion threatened Fort Pitt (now the city of Pittsburgh). The British commander and the leader of the local militia treated with a delegation of Delaware Indians, and, "out of our regard for them, we gave them two Blankets and a Handkerchief out of the Small Pox Hospital. I hope it will have the desired effect." Indeed, the Delaware tribe soon suffered from an epidemic of smallpox (CBWInfo, 2005).

Apparently independently of this action, other officers contrived to cause outbreaks among their Native adversaries. British commander in chief General Sir Jeffery Amherst corresponded with Colonel Henry Bouquet, who was leading a relief force to Fort Pitt, indicating that the use of smallpox as a weapon against the rebellion was receiving general consideration and asking, "Could it not be contrived to send the Small Pox among those disaffected tribes of Indians?" Bouquet replied, "I will try to inoculate the . . . with some blankets that may fall in their hands, and take care not to get the disease myself" (CBWInfo, 2005). Although Bouquet's own troops apparently suffered exposure and an outbreak began among them as well, Amherst informed him, "You will do well to try to inoculate the Indians by means of blankets, as well as to try every other method that can serve to extirpate this execrable race" (Sunshine Project, 2008).

The practice continued into the American Revolution. As the war began in April 1775, the British Army responded to a smallpox outbreak in Boston by variolating not only soldiers, but also fleeing civilians. The intention may have been humanitarian or a strategic effort to prevent a wartime epidemic, but it was also recognized that variolation made the recipient infectious for a period of days, and this could be advantageous if the refugees were headed for rebel camps. General George Washington therefore wrote that "the enemy intended spreading the Small pox amongst us." More concretely, other records do indicate the hope of causing outbreaks among the Continental Army. In a letter written to General Charles Cornwallis shortly before his surrender at Yorktown in 1781, General Alexander Leslie wrote that more than "700 Negroes are come down the river in the Small Pox" and that he planned to "distribute them about the Rebell plantations" (CBWInfo, 2005; Wheelis, in Geissler and Moon, 1999: 28–29).

Washington had already been confronted with the outbreak of a smallpox epidemic in January 1777. Inoculating the Continental Army meant that all recipients of the treatment would be incapacitated while they recovered, and a number could still be expected to contract the virus and die. He determined this approach to be less fraught than taking no precautions and ordered the inoculations to be performed secretly to avoid alerting the British. The approach succeeded, as few became ill and the smallpox death rate in the general population declined significantly as well (Marble, 2010).

Other manifestations of biotechnology in warfare continued into the nineteenth and twentieth centuries. The poisoning of wells continued from antiquity into modernity, with incidents reported during the American Civil War and South African Boer Wars, and in the 1990s in the ongoing civil strife between the Turkish government and the Kurdish Workers' Party (CBWInfo, 2005). Mau Mau rebels in 1950s Kenya poisoned British cattle using local plant extracts (Millet, in Wheelis, Rozsa, and Dando, 2006: 232). More recently, the Afghan Taliban employed mass poisoning against government forces. In 2013, seventeen police trainees were drugged with rat poison at a banquet and then shot as a precaution against recovery (Nordland, 2013).[9]

The practice of using bioweapons to remove native peoples from contested land has been documented in modern times as well. In the 1950s and 1960s,

> agents of Brazil's Indian Protective Service worked in collusion with land owners who wanted to expand their holdings by removing indigenous people from tribal areas. To do this, they turned to chemical agents, such as arsenic laced with sugar, and biological agents including measles, tuberculosis and influenza. Some Amazonian tribes were apparently deliberately variolated with live small pox rather than inoculated against it. More than 300 IPS officials were either dismissed or charged with criminal conduct. (Wheelis and Sugishima, in Wheelis, Rozsa, and Dando, 2006: 285–286)

The tradition of using biotechnology in armaments also continued. In the 1960s, the Vietcong enhanced the infectivity of their antipersonnel devices by smearing impaling *pungi* sticks with excrement (Roberts, 2003: 14). In the late nineteenth century, different major powers similarly had examined the possibility of contaminating bullets with fecal matter (Wheelis, in Geissler and Moon, 1999: 33). And glass shells capable of carrying biological and chemical toxins were devised during the Crimean and American Civil Wars (Murphy, 1985: 26–28).

THE ORIGINS OF STATE BIOWEAPONS PROGRAMS

Despite this evolution from hurled clay pots filled with serpents to modern artillery, the use of bioweapons remained ad hoc, often desperate, battlefield tactics until the interwar period of the twentieth century. Pathogens were now recognized as the source of disease transmission, and industrial production methods were available to raise and stockpile sufficient quantities of them for strategic purposes. However, widespread norms against WMD use had already been reflected in international laws prior to World War I, and the horrors of chemical weapons on the battlefield led to additional treaties prohibiting BW, even though they had negligible use during the conflict. Ironically, it was at this point that all of the major industrial states independently began their BW research and development programs.

Precursors to the Arms Race

Developments such as the introduction of picric acid–bearing shells into rapidly mechanizing mass armies provided the impetus for international conferences to attempt to limit the destructiveness of modern warfare. The 1874 Brussels Conference on the Laws and Customs of War prohibited "poison or poisoned weapons," while a subsequent declaration by the 1899 Hague Peace Conference foreswore "projectiles the object of which is the diffusion of asphyxiating or deleterious gases." A subsequent Hague conference in 1907 also banned projectiles containing poisons (Guillemin, 2005: 3).

These agreements clearly did little to prevent the deployment of the first chemical weapons attack by Germany at Ypres, Belgium, in 1915. Nor did they prevent it from practicing agrowarfare against Allied packhorses and mules shipped to its Entente foes from noncombatant states. German spies and saboteurs paid dockworkers to infect the animals with anthrax and glanders, resulting in entire shipments perishing and denying beasts of burden to the British and French armies when mechanized transport was still not widely available (Dando, 2006: 16; Smithson, 2011: 230). Guillemin (2005: 21) argues that "from the German perspective, this violated no international norm." However, Price (1997: 50) notes that the German military's *Kreigsraison* (doctrine of military necessity) specifically ruled out individual and mass poisoning as well as "the propagation of infectious diseases," despite justifying the use of chemical munitions and gas.

The Interwar Period: Treaties and Experiments

While the limited use of biological weapons attracted little attention, the deployment of chemical weapons in World War I produced a revulsion against weapons of mass destruction throughout the international arena, although many countries would later build stockpiles of such weapons prior to and during World War II. Nonetheless, numerous international efforts were made to eliminate both conventional and unconventional armaments during the next decade. Among other provisions, the 1922 Treaty of Washington banned the use in war of "asphyxiating, poisonous or other gases, and all analogous liquids, materials or devices," but the refusal of France to sign it and the United States to ratify it quickly led to efforts to secure an effective alternative (Guillemin, 2005: 4).

The result was the 1925 Protocol for the Prohibition of the Use in War of Asphyxiating, Poisonous or other Gases, and of Bacteriological Methods of Warfare (Klotz and Sylvester, 2009: 46). More commonly known as the Geneva Protocol, the agreement was intended to ban first use of chemical weapons, although it proscribed neither possession of them nor counterattacks. At the insistence of Poland, bacteriological weapons were added as well (Price, 1997: 111). This concern over what was still essentially an unfielded category of modern weapon was possibly due to the Polish experience of the 1919–1921 Soviet-Polish War. By the time of Geneva, a Soviet BW program had already been initiated, with experiments in cultivation and dissemination by "scientists [who] attached crop sprayers to low-flying planes" (Alibek, 1999: 6).

Despite signing the Geneva Protocol, the Soviet program continued, and all other major powers of the time that ratified it soon were conducting their own BW research and defensive programs.[10] Part of the motivation to do so stemmed from the suspicion that rivals were also secretly developing BW programs. The available evidence suggests that, as with other arms races, a security dilemma was at work: research into defenses against biowarfare necessitated first developing and testing BW delivery systems, making it impossible to determine whether offensive intent hid behind the programs.

Paradoxically, the new international norms against warfare may also have played a role. Some interwar military theorists believed that bacteriological warfare would be more civilized than other forms of combat because it could be accomplished without physical violence against the human body or widespread damage to infrastructure. Another source of appeal was a fascination with modernity: "The emerging research fused the expanding science of infectious diseases with emerging technologies of aerial warfare." In the 1920s, military planners argued that biological weapons had little battlefield utility because they were hard to control, slow to act, and unpredictable in their effects, but that they held promise for surprise aerial dissemination against civilian industrial targets (Guillemin, 2005: 6–7, 11).

Other developments were also spurring an interest in the destructive capability of pathogenic agents. World War I and its ensuing conditions of widespread refugees, famine, and lack of adequate sanitation led to casualties that ultimately dwarfed even the terrible toll produced directly by combat (approximately ten million combatants and seven million civilians). The ten million deaths attributed to typhus and famine in the Russian Civil War reportedly persuaded the Kremlin to pursue its BW program (Alibek, 1999: 32).

And the 1918–1920 influenza pandemic is estimated to have killed approximately fifty million people worldwide, a figure that approaches the global number of deaths attributable to World War II. Although it was known as the "Spanish flu" because it received intensive media coverage after Spain's King Alphonso XIII contracted it, evidence suggests that the strain first appeared in the United States, in Haskell County, Kansas, in March 1918. Army draftees were sent from there to Fort Riley, where so many were being mobilized that the Army chief of staff billeted them with only half the amount of space per person recommended by the surgeon general to prevent the transmission of disease. Similarly crowded hospitals, trains, and troop ships allowed the spread of the intensely virulent strain, first to the East Coast of the United States and then to France. Unusually, as it seemed to observers at the time, the victims were disproportionately otherwise healthy young males (Marble, 2010). As with the Mongol and Spanish Empires, the war machine of the emerging American imperium unwittingly contributed to a pandemic that devastated entire nations under its hegemony.[11]

Since the first state BW programs began in the 1920s, approximately forty agents have received extensive testing. These programs were always constrained by a combination of norms, fear of retaliation, and technical difficulties. Most stocks did not remain potent for a long time and required particularly favorable conditions for reproduction. In the case of anthrax, for example, one kilogram could kill thousands. However, a missile bearing the bacteria will not be effective using a high airburst or by embedding itself in the ground. Too much depended on the discretion of the wind, and it was impossible to adequately protect combatants and civilians, at least not without alerting enemy troops. "The early programs were started without much proof that any pathogens could be made into effective weapons that a modern military could use with confidence. It took authoritative scientists to convince civil and military leaders that these unusual weapons could have future value" (Cohen, 1998; Guillemin, 2005: ix, 15).

The French BW Program

Jeanne Guillemin (2005: 11, 15, 24–25) identifies France as having initiated the first state BW program, one that began in 1921 under the command of Auguste Trillat, director of the Naval Chemical Research Laboratory, and that continued until German forces overran the country in 1940. Reportedly, during a French visit to a German pharmaceutical plant in 1919 conducted under the inspection terms of the Treaty of Versailles, the plant director revealed that Germany was continuing its wartime work on bacteriological agents. Concerned by this development, and with an abundance of bacteriological research facilities in the country as a result of the legacy of Louis Pasteur, the French initiated a military program that conducted numerous trial disseminations of liquid cultures in bombs and artillery and by airborne aerosol spray, as well as tests on animal subjects at the Army Veterinary Research Laboratory in Paris. Trillat viewed such slurries (liquid concoctions of pathogens and nutrients) as the best medium for maintaining the virulence of the select agents, although later researchers demonstrated that bacteria could be kept viable as bioweapons by freeze-drying them as well (Guillemin, 2005: 25).

At the time, Pasteur Institute director Emile Roux claimed that "French scientists do not attack, but they study means of self-defense" (Lepick, in Geissler and Moon, 1999: 71). However, as with subsequent BW programs and defense research, it soon became difficult to separate offensive characteristics of the research. Trillat argued that tactical BW would be ineffective against combat troops, but strategies for targeting civilian populations and industries should be explored. France in the 1930s considered a number of anti-bioweapon civil defense or homeland security measures, including aerosol dispersal of antiseptics; monitoring air, dust, water, and food for pathogens; and providing vaccinations and face masks to citizens. However, Trillat argued that it would be impractical to attempt to provide adequate defenses against any possible attack and favored instead deterrence through the threat of a symmetrical response (Guillemin, 2005: 25), the same strategy that would form the basis for the nuclear balance of terror between superpowers during the Cold War.

The German BW Program

French BW planners justified their efforts by fears of a secret German program conducted in defiance of the Treaty of Versailles. In actuality, the German military was concerned by reports of Soviet bacteriological bombs but was skeptical about their potential effectiveness and feared blowback against their own forces if they attempted to deploy bioweapons themselves. By 1942, British intelligence had discovered that German troops were indeed training with biological and chemical weapons but had a no-first-use doctrine (Guillemin, 2005: 24, 26, 41–42).[12]

Records indicate that Germany feared that even its conventional blitz would provoke Britain into using BW in response. Perhaps uniquely among great powers in WMD arms races, Germany initially decided that it should not even possess the armaments for fear of initiating a BW in-kind response. Therefore the German program, which did not begin until 1941 after discovery of the extent of the French program after the fall of Paris, did not enjoy

Hitler's support and only developed biodefense technology and not even BW for retaliatory purposes (Geissler, in Geissler and Moon, 1999: 91–125). [13]

The Prewar Soviet BW Program

Just as French BW efforts were spurred by fear of German capabilities, and the German program by concern over Soviet advances, the Kremlin worried about its once and future antagonists, as well as the possibility of internal threats. The Soviet Revolutionary Military Council ordered the development of typhus weapons in 1927 and placed them under the control of military intelligence. By the 1930s, the Soviets were harvesting the pathogen, which kills approximately 40 percent of its victims, by injecting the bacteria into chicken embryos grown in industrial egg farms, or from the liquefied remains of infected rats. The bacteria could be disseminated either as powder or in an aerosolized form.

In the 1920s, the KGB opened Laboratory 12 to develop substances that could be used for quick and efficient assassinations. Out of this program grew Project Flute, which produced paralytic or fatal "psychotropic and neurotropic biological agents for use by the KGB in special operations." It also reportedly developed plague bacteria in a powder form in the 1940s that was planned for use against Marshall Josef Tito of Yugoslavia, and the ricin used to assassinate Bulgarian dissident Georgi Markov in London in 1978.

By the mid-1930s, Soviet scientists were experimenting with an array of select agents that included Q fever and glanders. Owing perhaps to poor safety conditions, possibly hundreds of scientists and lab workers became infected by their specimens (Alibek, 1999: 32–35, 172, 173). Even though it too adopted the Geneva Protocol, in 1928, the Soviet Union at the same time ramped up its efforts and created the largest BW research program in the world during the interwar years, a status it would maintain in the decades beyond (Rimmington, in Wright, 2002: 106).

THE WORLD WAR II FERMENTER

Given the growing amount of BW research during the interwar period, a peacetime era supposedly governed by the regime of the Geneva Protocol, it would perhaps have been reasonable to expect large-scale deployment of bioweapons when the international order disintegrated during World War II. That they were scarcely used, if at all, in interstate warfare is a testament not only to the universal normative antipathy that had developed toward such weapons, but also to fear of retaliation. This is not to say, however, that BW programs did not advance substantially during the war years. Indeed, one belligerent in particular developed extensive research from its use of bioweapons against noncombatants that would lay the foundation for all postwar state programs.

Japan began testing a variety of toxins (including plague, typhoid, hemorrhagic fever, and gangrene) on thousands of POWs in Manchuria. During the course of the war, over two thousand Soviet and Chinese human guinea pigs were killed by these infections. The British conducted a test anthrax airburst over domestic sheep in 1942, and the United States and Canada began work on five-hundred-pound brucellosis cluster bombs in 1943 (Murphy, 1985: 28–31).

Other methods for transmitting bacteriological weapons relied on the use of living delivery systems. Japan bred fleas carrying plague vectors that were deposited in Chinese villages. Great Britain developed, but never deployed, "anthrax cakes" to feed to German cattle. And suspicious outbreaks of tularemia at Stalingrad in 1942, and of Q fever in the Crimea in 1943, rapidly incapacitated German troops on the eastern front (Preston, 2009: 206, 225, 315). While no confirmed battlefield uses of bioweapons took place during the most destructive war in history, all of the belligerents had BW programs, and the conflict was nonetheless responsible for the advent of massive state bioweapons production, as well as research and development—ostensibly purely for defensive purposes—that continues to this day.

Another World War II development that continues to feature in twenty-first-century military research was the strategic mass deployment of narcotics to troops for offensive force projection. During World War II, American troops were issued amphetamines in their field kits for alertness, and GIs later created the speedball by adding heroin to the mix during the Korean War (Rexton Kan, 2008). Fighters in the Syrian civil war have both trafficked in and become addicts of an amphetamine called Captagon, originally created in the West in the 1960s as an antidepressant, which produces a sense of reckless abandon that keeps fighters from sleeping for days and permits remorseless killing and disregard of being tortured (Holley, 2015).

The German use of narcotics in World War II signifies both an early recognition of the strategic uses of troop enhancement and the risks created by substance abuse: Pervitin, developed in Germany in the late 1930s and initially available legally as a recreational stimulant, was issued to troops during the early days of blitzkrieg so that they could go for days without sleeping. Although successful with first-time users, it was no longer making a difference in performance by the time it was tried during the devastating defeat at Stalingrad in 1943. Today, Pervitin is better known by its street name, crystal meth. At the same time, Hitler's physician reported prescribing him with steroids and Eukodal, which apparently kept him euphoric and confident of victory even as his strategic position deteriorated (Breitenbach, 2015).

The Japanese BW Program

Japan was the only party in World War II or in modern history known to use biological attacks with bacteriological weapons on a wide scale. Despite its meteoric rise as a regional power that began half a century earlier, imperial Japan before and during World War II remained resource poor by the standards of Europe, and it remained open to asymmetric strategies to compensate (for example, ordering kamikaze suicide attacks against American vessels and launching an armada of balloon-carried explosives to be carried across the Pacific). Japanese development of bacteriological weapons may have begun as early as 1936 (Murphy, 1985: 28), and well-documented efforts to spread disease in occupied China and Manchuria under the guise of natural outbreaks occurred during 1939–1945 (Guillemin, 2005: vii). The father of Japan's BW program, General Shiro Ishii, reportedly became interested in bioweapons when he concluded that they must be effective or else the international community would not have bothered developing restrictions against their use (Klotz and Sylvester, 2009: 47).

Ishii's three-thousand-member-strong program team, officially the Kempeitai Political Department and Epidemic Prevention Research Laboratory, but generally known as Unit 731, conducted human experiments that killed several thousand Chinese subjects in camps and laboratories, and perhaps several hundred thousand civilians in the field. In 1940, Ishii's command dropped grain and rags contaminated with plague from aircraft over cities, as well as small porcelain containers filled with plague-infected fleas. Inhabitants of the affected areas blamed the persistence of plague-infected insects for high numbers of fatalities for years after the war (Klotz and Sylvester, 2009: 46, 48). Other outbreaks caused by intentional release of biological agents included cholera and typhoid (Dando, 2006: 24). The Japanese Army was unable to determine precise numbers of casualties (it estimated one thousand killed and another one thousand sickened shortly after the aerial attacks), and it separately provided contaminated food to the civilian population as well (Guillemin, 2005: 84). Subsequent estimates have placed the number killed as high as four hundred thousand (Sandberg, 2014). Similar to Germany using chemical weapons in concentration camps but not on the battlefield, Japan's military leadership mostly resisted calls for biological attacks in the Pacific.

An exception occurred in August 1939, when the Red Army routed the Japanese on the Mongolian-Manchurian border, and Unit 731 left behind a suicide detachment to infect the Khalkin Gol's river water with a variety of pathogens. It is unclear whether these had any effect on the Soviet soldiers. But in 1942, at least a thousand Japanese soldiers died from blowback during an attempted contamination effort, leading to Ishii's removal as head of Unit 731 (Guillemin, 2005: 84, 85).

Despite this, Japan's human-subject BW program continued and left a legacy that endured well beyond the end of the war. The research ultimately contributed to the production of effective vaccines and treatments for plague, typhoid, dysentery, and other diseases, but also to enhancing the BW programs of the powers that obtained the Japanese data. In 1945, the Office of Strategic Services (the forerunner of the Central Intelligence Agency) of the United States determined that Japan had the most accomplished BW program in the world. It held data on disease progression and pharmaceutical response that the Allies could not similarly obtain through human experimentation, unique autopsy records, and scientists whose expertise would be used against the United States if they cooperated with its enemies. Despite the fact that these researchers had conducted human experiments on par with those of the German scientists that it branded as war criminals, the United States did not seek any charges against the scientists and officers of Unit 731, instead offering them amnesty in exchange for their cooperation. The Americans concluded that only anthrax and plague had been effective bioweapons, and particularly welcomed the data on inhalation anthrax (Guillemin, 2005: 76, 77, 79, 86).[14] As with post-Soviet WMD researchers fifty years later, the solution to the threat of proliferation pursued by the United States was to reemploy the scientists.

British Commonwealth BW Programs

By contrast, and despite alarm about a reported German program, the British BW program did not begin in earnest until the nation was already under direct military threat. Winston Churchill had written in 1925 about foreseeing the introduction of "Blight to destroy crops, Anthrax to slay horses and cattle, Plague to poison not armies but whole districts," but yet the program evolved through "inadvertent escalation."

The Ministry of Defence established the Subcommittee on Bacteriological Warfare in 1936. By 1940, with its very existence challenged by the Third Reich that had predicted such a move, bioweapons research was added to the already operational chemical weapons program. Ultimately, at a chemical weapons site at Porton Down, numerous tests of select agents were conducted that indicated that, for all of their potential drawbacks, biological weapons were actually between one hundred and one thousand times more potent than the deadliest chemical armaments. One program that grew out of this research was Operation Vegetarian, which—although they were never deployed—produced five million linseed cakes laced with *B. anthracis*, a fraction of which would have wrought long-term devastation on the German cattle industry (Guillemin, 2005: 12, 40–55). At least one cake continued to yield virulent cultures a decade after it was produced (Dando, 2006: 27).

A key figure in the development of both the British BW program and one in his native Canada was Frederick Banting, who had previously won a Nobel Prize for discovering insulin as a treatment for diabetes. Banting, who feared a German biological attack against drinking water supplies, argued that pathogens could be freeze-dried and still maintain their virulence, and that this would be the best medium for effective aerosol dissemination. He even considered sending *B. anthracis* spores in this fashion through the mail to reach targets. He oversaw Allied tests in Canada involving blow-drying typhoid bacteria onto sawdust and then dispersing it from aircraft (Guillemin, 2005: 46, 48, 52).

The American Wartime BW Program

The United States initiated its BW program in 1941, and its industrial might soon became vital to meeting the bioweapon production needs of its Anglophone allies. In 1942, the War Research Service was established to manage military research and development, including the BW program. Its scientists examined seventy potential pathogens, both lethal select agents and others that would only incapacitate with illness.[15] Among them, *B. anthracis* in its dormant spore form ranked as the most important agent, but researchers looked beyond anthrax to work as well with brucellosis, tularemia (rabbit fever), and agricultural pathogens like fungi and insects (Guillemin, 2005: 7–8, 27–28, 34, 36, 59).

With the recognition that the quantity of pathogens needed for arsenals of bioweapons would be immense, the United States began an industrial-scale manufacturing program of biological agents and vaccines. The virulence of select agents was increased by serially passing samples through animal hosts.[16] When the British ordered half a million anthrax munitions from the United States in 1943, Churchill described it as a "first installment" (Guillemin, 2005: 66–68, 74).

Despite the large scale of the American BW program, the United States did not deploy either biological or chemical weapons at any point during the war. In part, this was because American military strategists believed that BW would be ineffective against frontline troops because of the lag between exposure and symptoms, and also because vaccinations were already available against many of the weapons. As interwar European planners had, they conceived of BW as being most potentially useful as terror weapons employed against civilian populations in industrial production areas. Ultimately, however, this objective was accomplished by conventional strategic bombing, and finally with the first nuclear weapons. And just as some Manhattan Project scientists proposed turning over American nuclear technology to the new United Nations to avoid creating the incentives for other states to engage in arms races, so too did some BW scientists suggest doing this with

American bioweapons, or at least making information about the full extent of the program public (Guillemin, 2005: 28, 30, 37, 74). That any of these proposals were taken seriously by the US military seems unlikely, and the rapid emergence of a rival power at war's end ensured that both BW secrecy and production would continue for decades.

The Soviet Wartime BW Program

Most descriptions of the Soviet BW programs and its accomplishments from the Russian Civil War to the fall of communism were based on the accounts of two lead Soviet bioweaponeers, Vladimir Pasechnik and Ken Alibek, who defected to the United Kingdom and United States, respectively. Much of Alibek's frightening account of the largest-scale BW program in history and its engineered, untreatable deadly viruses is (at least publicly) uncorroborated, and some elements are contested. Perhaps most controversially, Alibek claimed in his memoir to have discovered as a graduate student that the desperate Red Army unleashed bioweapons against invading German forces.

By his account, a wave of tularemia hit the panzer divisions besieging Stalingrad in 1942 and then spread to the Soviet forces and civilians; initially all of the victims were German. Ultimately, one hundred thousand Soviets were sickened and sought treatment, a figure ten times higher than the numbers of tularemia victims in preceding or previous years, and 70 percent of cases were the pneumonic form of the disease, "which could only have been caused by purposeful dissemination." Then, an outbreak of Q fever, which had otherwise never been reported before in the Soviet Union, afflicted German forces in the Crimea in 1943. Alibek reported that he later received confirmation from a lieutenant colonel who worked on the wartime Soviet BW program that the lesson had been learned that anything that could be sprayed at the enemy could blow back, and that production needed to be moved farther away from the front lines, which is why the program was set up on Vozrozhdeniye ("Rebirth") Island on the Kazakhstan-Uzbekistan border (Alibek, 1999: 30–31, 36).

When the Soviets overran Manchuria, they obtained the records of Unit 731 and were shocked to realize that Japan had a more advanced BW program. The Soviet Union attempted to use as international propaganda the fact that the United States was now employing Japanese war criminals who had experimented on Chinese civilians and American, British, and Commonwealth POWs, but it quickly moved to incorporate the captured Japanese blueprints into its own BW facility designs (Alibek, 1999: 37).

THE COLD WAR APOGEE

With both of the superpowers that emerged from the ashes of World War II supremely distrustful of the other and intent on possessing the most versatile and formidable arsenal available, state BW programs continued throughout the forty-year Cold War. In the middle of this period, a new binding international agreement to ban all biological weapons was introduced by the United States and entered into effect, but, as it permitted research for defensive purposes, BW development continued on both sides. There is little evidence that either superpower employed bioweapons against their regional adversaries in their limited wars during this period; instead, it was their own citizens who felt the effects of the research programs.

The United States and Bioweapons

At the outset of the Cold War, American military strategists believed that anthrax bombs had greater strategic utility than atomic bombs, and for the next twenty years the goal of their BW program was to develop bioweapons with the same magnitude of offensive capability as nuclear weapons. Unlike the atomic bomb, which generated a great deal of public concern, the secrecy surrounding the wartime BW program and the integration of Japanese data was not lifted, and bioweapons were quietly incorporated into the Cold War strategic arsenal. A 1951 report issued by the Joint Chiefs of Staff argued that bioweapons were a sound investment for the US military because they were far less expensive than nuclear arms while being potentially as destructive (Guillemin, 2005: 12, 90, 92, 93). Subsequently, the United States mass-produced Q fever and equine virus beginning in the 1950s by injecting hundreds of thousands of chicken eggs on conveyor belts with pathogens, which then naturally incubated, using technology widely available today for commercial purposes (Preston, 2009: 207).

In 1956, the United States changed *The Law of Land Warfare* Army doctrine manual to remove references to purely defensive use of bioweapons, noting instead that they could be used to enhance military effectiveness if the president so determined. President Kennedy subsequently supported a robust BW program as part of his doctrine of flexible response that was designed to provide greater leeway than Eisenhower's nuclear brinksmanship (Guillemin, 2005: 107, 113, 114).

By the end of the decade, however, President Nixon believed that he could generate renewed international support for the United States after opprobrium for using chemical defoliants in Vietnam by curtailing its chemical and biological weapons programs, which by this point Washington dismissed as redundant to its strategic nuclear capability. Nixon ordered a unilateral halt to BW production in 1969 and limited research to purely "defensive" capabilities. He would subsequently propose a complete international ban on biological weapons, including both pathogens and toxins, ultimately enacted as the Biological Weapons and Toxins Convention (BWC) (Guillemin, 2005: 123, 124, 126).

By this juncture, the United States had developed a bioweapons arsenal that included a variety of bacterial and viral select agents targeting humans, and fungi and plant pathogens to be used for agrowarfare. Cobra venom and other toxins were also harnessed for use in clandestine intelligence operations (Roberts, 2003: 16, 46). And munitions had been developed that could be filled with pathogens (Kortepeter and Parker, 1999: 524).

While these agents were never employed against foreign troops or civilians, it was a different story in terms of domestic testing. In 1946, the US Navy announced that its research unit had tested bubonic plague on inmates at San Quentin Prison. Refusals by enlisted men to serve as test subjects for tularemia treatment experiments—culminating in a sit-down strike—led the surgeon general to persuade the leadership of the Seventh-day Adventist Church to provide volunteers instead. Between 1955 and 1957, nearly two hundred participants joined Project Whitecoat for inhalation and injection experiments, with a number of participants becoming incapacitated for weeks despite the administration of antibiotics. Other experiments tested subjects for exposure to Venezuelan equine encephalitis, cutaneous anthrax, and exotic tropical diseases. Experiment records indicate that all test subjects survived in good health, but that they believed that they were doing purely defensive research, not participating in weapons programs. Despite the Nixon initiative, these tests continued until 1976, one year after the United States ratified the BWC (Guillemin, 2005: 95, 105, 106).

The BWC (discussed in greater detail later in this chapter) banned the development of bioweapons but permitted defensive research to continue. As was well recognized by the 1970s, this was such a big loophole as to render the agreement largely meaningless: preparation of defenses to biological attack requires research with the pathogens and their dispersal systems. Unit 731 had formally been engaged in primarily defensive research (Dando, 2006: 22).

The lack of any mechanism for monitoring or enforcement was also a serious flaw. Soviet scientists continued in their same capacities, albeit with formally revised job descriptions, after the Soviet Union signed the BWC (Guillemin, 2011: 2). And while the US Army Medical Research Institute for Infectious Diseases (USAMRIID) was formally only conducting defensive research, it paid "disproportionate attention to exotic biological agents and toxins, some of which are not even considered legitimate threats by the military's own intelligence data" (Reed and Shulman, in Wright, 2002: 63–64).

In the 1990s, at least three military research experiments were conducted that, while justifiable as defensive research, nonetheless highlight the fine line between offensive and defensive biotechnology. Project Jefferson sought to test whether commercially available anthrax vaccines would be effective against strains engineered by the Soviets to be drug resistant. Researchers did so by replicating Soviet efforts to create drug-resistant bacteria. Under Project Bacchus, researchers built a pathogen production facility with commercially available material to determine the ease with which non-state actors might be able to do the same. And Project Clear Vision built a dispersal bomb from Soviet schematics to test its effectiveness (Klotz and Sylvester, 2009: 88).

The Soviet Union and Bioweapons

The Soviet program, Biopreparat, which involved participation across government agencies including the Ministries of Defense, Health, and Agriculture, the KGB, and a web of academic and industry research labs, "stockpiled hundreds of tons of anthrax and dozens of tons of plague and smallpox." The development of delivery systems capable of disseminating bioweapons also advanced far beyond the crop dusters of the 1920s, with bomber fleets added in the 1940s and intercontinental ballistic missiles and warheads by the 1970s. Production increased to keep pace with the capacity of the delivery systems: "A single twenty-ton fermenter working at full capacity could produce enough spores to fill one missile in one or two days." And by the time the Soviet Union collapsed, efforts were underway to build the capacity to fill the numerous multiple independently targetable reentry vehicles (MIRVs) that could deploy from a single missile.[17] While the United States reportedly had two anthrax production specialists, the Soviet Union had two thousand (Alibek, 1999: x, 5, 6, 230). Ben Ouagrham-Gormley (2014: 37) notes that the Soviet program may have been overall ten times larger than the US program, and it continued its offensive research for more than twenty years longer, but its greatest hallmark would appear to be its monumental scale rather than the realization of its more innovative BW ambitions.

Alibek states that while the United States for safety reasons only tested agents that could be countered by antibiotics or vaccines, the Kremlin decided that the most useful pathogens were those against which no treatment was available, and that anytime progress was made against one of the agents, Biopreparat scientists were set to work making it even more resistant. Soviet-sponsored scientific and trade organizations across the globe sought out the most exotic and deadly pathogens and substances that they could obtain, to the extent that the KGB was nicknamed "Capture Agency One." Under its cover as a pharmaceutical and medical research company, Biopreparat was successful in obtaining samples from the West, including the Marburg virus, a more lethal cousin of Ebola, and in experiments with HIV, which was found unsuitable for military purposes given its long latency period. Other forms of biotechnology in the arsenal included psychotropic agents, toxins, and (separate from the small, internationally recognized Soviet stock) "tons of smallpox in our secret lab in Zagorsk," half an hour outside of Moscow (Alibek, 1999: 8, 18–20).

Soviet bioweaponeers began smallpox research in 1947 and made great strides after isolating a virulent strain brought to Moscow by a traveler from India in 1959 who infected dozens and forced a quarantine. After this incident, Soviet medical teams traveled to India with the expressed goal of smallpox eradication, but the KGB agents among them sent back the most promising samples for weaponization. In the 1970s, the military deemed smallpox so vital to its strategic interests that an order was issued to the BW program to maintain a stockpile of twenty tons of it at the ready. According to Alibek, after the World Health Organization announced in 1980 that the disease had finally been eradicated across the world, research on illicitly held stocks only intensified (Alibek, 1999: 59, 111, 112).

Other projects encompassed work begun in the 1950s on anti-livestock and anticrop weapons. These included foot-and-mouth disease and rinderpest for use against cattle, African swine fever for pigs, and other agents for use against chickens. The intent behind these transmissible select agents was that the targeted animals would pass along the pathogens and devastate large agricultural areas. As with smallpox, production facilities were hidden in plain sight in the middle of large urban centers (Alibek, 1999: 37, 38).

The degree of secrecy surrounding Biopreparat was such that many Kremlin officials plausibly denied knowledge of the full range of its activities. Boris Yeltsin, serving as governor of Sverdlovsk at the time of an anthrax outbreak there in 1979 that claimed an unconfirmed number of casualties,[18] was until that point unaware of the presence of a biological weapons facility in the region (Koblentz, 2009: 67, 114). The Sverdlovsk facility, operational since the 1940s, produced an outbreak when an air filter was not reinstalled properly after routine maintenance. Even the resultant contamination provided data: older victims were more vulnerable to anthrax, while many younger people in the affected zone did not even test positive for the spores. When the Soviets finally announced an anthrax outbreak—which they insisted was the result of black-market meat—twenty thousand of the fifty thousand intended recipients of vaccinations did not show up to receive them because of warnings of side effects. Similarly, a smallpox outbreak, near a Biopreparat facility in Kazakhstan in 1972, was described thirty years later by a public health minister as having been caused by the military (Guillemin, 2005: 141–143).

Despite these instances, as with the Americans, there is no definite evidence that the Soviets ever practiced biowarfare. Unusual outbreaks of glanders were reported in Afghanistan in 1982 and 1984 (Preston, 2009: 225), and other reports claim that the Red Army sprayed mujahidin with fungal mycotoxins, so-called yellow rain (Murphy, 1985: 39–49). Alibek (1999: 268) reported hearing of an aerially disseminated glanders attack that corresponded to one of the Afghan outbreaks, and also of the use of an "antimachinery" corrosive biological weapon. Investigations into allegations of yellow rain in Southeast Asia remain contested, with some investigators finding that the substances were actually waste products from bees, while others have insisted that reported outbreak symptoms and victim interviews were consistent with exposure to a mycotoxin (Meselson and Robinson; Katz, in Clunan, Lavoy, and Martin, 2008: 72–115).

The Soviet Union had ratified the BWC, trusting that the lack of verification mechanisms would ensure the continuance of its research program. Biopreparat activities actually peaked in the late 1980s, with what Alibek reports were sixty thousand researchers in the BW program and President Mikhail Gorbachev increasing its budget despite ostensibly thawing relations with the West.[19] Mobile bioweapons labs, which the United States would later accuse Iraq of employing to escape international inspectors, were devised for that purpose by the Soviets in 1988. The British and Americans were aware by this point of the scale of Soviet efforts but did not press for the elimination of BW programs because they believed that it was a greater priority to win concessions from Gorbachev on nuclear arms. The Kremlin responded to the scrutiny by planning to cut production of bioweapons and rely more heavily on research and stockpiling by dummy civilian companies (Alibek, 1999: 145, 146, 152, 178).

The collapse of the Soviet Union brought a halt—at least officially—to the work of Biopreparat. In 1992, Russian Federation president Boris Yeltsin signed a decree banning offensive biological research and cutting defense research budgets by half. Open-air testing at Rebirth Island halted, and Russia, the United States, and the United Kingdom agreed to convert all bioweapons facilities into centers for peaceful scientific research.

However, offensive biotech development appears to have continued not only in Russia but possibly in other former Soviet republics as well. And some of the large number of Biopreparat personnel appear to have sought work elsewhere. Some of them have been hired by other governments (more details later in this chapter), but others may have been in the employ of the Russian mafia, which is known to have hired former KGB officers. Alibek makes the ominous connection that, in 1995, the anticorruption chairman of the Russian Business Roundtable and his secretary died suddenly of a fast-acting and mysterious illness (Alibek, 1999: 28, 176, 245, 257).

OTHER STATE PROGRAMS

The end of the Cold War and a period of internal weakness in Russia did not mean the end of international concern over potential biotech threats. Defensive research requiring the maintenance of stockpiles of pathogens continued in France, Russia, Germany, the United States, Canada, the United Kingdom, and Japan, and BW programs were also documented in South Africa and Iraq. Suspicion exists that programs were developed as well in Egypt, North Korea, Israel, and Rhodesia (now Zimbabwe) (Congressional Research Service, 2007: 11; Guillemin, 2005: 151–152). Syria tacitly acknowledged possessing unspecified BW when it declared its chemical weapons in 2013 (Warrick, 2013). Alibek (1999: 274) notes that Cuba has accused the United States more than a dozen times of initiating biological attacks against it, but that, particularly given the island nation's advanced medical sector, it probably has its own BW program.

China and Bioweapons?

Additionally, the United States has suspected the People's Republic of China of developing bioweapons as well:

> The United States believes that China began its offensive BW program in the 1950s and continued its program throughout the Cold War, even after China acceded to the BWC in 1984. Undoubtedly China perceived a threat from the BW programs of its neighbor, the Soviet Union. (US Arms Control and Disarmament Agency, 2005)

By 2015, the United States regarded China to be in full compliance with BWC while noting that it engaged in "biological activities with potential dual-use applications" (US Arms Control and Disarmament Agency, 2015).

As will be detailed in subsequent sections and chapters, the United States has sanctioned Chinese firms for trafficking in material for use in BW programs, and, despite protests of innocence by Beijing, individual Chinese military scientists have gone on record claiming to be working on advanced genetic weapons and insect-borne pathogens.

The North Korean BW Program

Reports indicate that the Democratic People's Republic of Korea has an active BW program, with at least twenty production facilities modeled on Biopreparat. The program reputedly is weaponizing a number of pathogens, including pneumonic plague and cholera, and has large stores of *B. anthracis*.

According to defectors, South Korean intelligence agencies and other sources, the nation's Fifth Machine Industry Bureau has led a successful effort to build one of the world's most extensive biochemical warfare programs. The weaponry is thought to have the potential to decimate North Korea's southern neighbor and the 28,000 U.S. troops stationed there, and to disrupt the regional economy. The gravest danger may be that North Korean dictator Kim Jong Il could sell his weapons to terrorists. (Cooper, 2009)

The South African BW Program

There are no uncertainties concerning South Africa's experiments with military uses of biotechnology, which occurred in the context of the unique strategic dilemmas confronted by the minority white apartheid regime. The government of President P. K. Botha perceived an existential threat from both internal insurgents and neighboring postcolonial black governments and determined that these warranted unconventional approaches to defense.

As a response, in 1981 it launched Project Coast, which worked with pathogens, including Marburg and Ebola, and hallucinogens and sedatives that could be used to kill political prisoners without visible signs of violence. Project Coast also looked into the feasibility of weapons that would only be effective against blacks, including a stealth contraceptive. Although managed by the Defense Department, Project Coast research was conducted by universities and private companies. After the end of white rule, project head Wouter Basson was employed by Libya as a consultant and is also known to have had contact with Iran and with racist militia groups in the United States. Washington, fearing that Basson's expertise would proliferate to rogue actors, requested that the new government of Nelson Mandela rehire him (Guillemin, 2005: 154–156).

The South African security apparatus is documented to have employed biological weapons on multiple occasions throughout the 1980s. The Civil Cooperation Bureau, which performed clandestine operations against the enemies of the regime, is known to have used poison against African National Congress activists. Among its operations revealed after the end of the apartheid era were the assassination of domestic opposition figures (Truth Commission, 2001) and, in 1989, the contamination with cholera of the water supply for a refugee camp in Namibia (Committee, 2006: 44).

Basson aside, many Project Coast employees did not have a clear understanding of the political and security implications of their research. Many were recruited ostensibly for biodefense projects, having been told that the Soviets were providing bioweapons to insurgents in Angola and Mozambique, or even on the pretext that they would be helping the pharmaceutical industry deal with overpopulation (Klotz and Sylvester, 2009: 13, 52–55).

The Iraqi BW Program

Iraqi deputy prime minister Tariq Aziz best summed up the Saddam Hussein regime's moral ambivalence toward WMD in 1988, when he justified the use of chemical weapons against Iranian forces by claiming, "There are different views on this matter from different angles. You are living on a civilized, peaceful continent" (Price, 1997: 142).

Following its ejection from its occupation of Kuwait, while Iraq fully cooperated with the United Nations Special Commission (UNSCOM) on eliminating its chemical weapons (CW) arsenal, it remained deliberately vague about its BW capabilities. Despite Hussein bragging in 1990 to other Arab leaders that he had a biological deterrent that would stop the United States in Operation Desert Storm, by mid-1991 the regime was repeatedly telling UNSCOM that it had no BW program. It next chose to surrender its least-modern BW stockpiles while hiding its most effective agents. This approach was the obverse of Hussein's prewar strategy in that he calculated that he could not afford to look weak domestically or to regional rivals by appearing to have been disarmed of his most potent weapons (Smithson, 2011: 14–16, 28–30).

Ultimately the regime admitted that it possessed a BW program despite being a signatory to the BWC (Kortepeter and Parker, 1999: 523). By Baghdad's own admission, it had sufficient agents to produce 8,400 liters of anthrax, 19,000 liters of botulinum, and 2,000 liters each of aflatoxin and clostridium (Roberts, 2003: 20–21). Lacking the means to fill warheads with freeze-dried pathogens in powder form as the superpowers had, Iraq filled SCUD missiles with slurry that would not have disseminated far upon impact (Klotz and Sylvester, 2009: 65).

After acknowledging the existence of its program in 1995, Iraq announced that it had destroyed all specimens, a claim that appears to have been confirmed after the American invasion in 2003. The decision to create an Iraqi BW program was made in 1974, immediately after the introduction of the BWC, and Western-trained Iraqi scientists began research in 1984 and production in 1988 with the established catalog of pathogens, as well as camel pox and fungal mycotoxins (Guillemin, 2005: 153–154). But, having deemed them impractical, it had never built the mobile labs that the United States accused it of using to conceal its BW program and supplied to the United Nations as a justification for war (Vogel, 2012: 136).

THE BIOLOGICAL WEAPONS CONVENTION

The Iraqi case, along with that of the Soviet Union, provides evidence that, as with the 1925 Geneva Protocol, the international regime governing bioweapon arms control actually encouraged the most egregious offenders to intensify their BW development programs. Approaching the half centenary of the BWC, it is worth examining how it has impacted the international norm against bioweapon usage in the decades since its inception.

Trust, Not Verify?

As noted earlier in this chapter, the Nixon administration determined that the United States could regain international goodwill that was being lost over the Vietnam War by curtailing its BW program, which it determined to be superfluous to its strategic nuclear arsenal. Even without a positive international reception for his proposal for a global ban on bioweapons, Nixon may have acted unilaterally in this regard because he did not trust in the effectiveness of arms control treaties. Given that the Geneva Protocol had not prevented the buildup of state bioweapons industries, this view held some merit. However, the protocol only banned the use of bacteriological weapons, not their manufacture or research. The new international agreement would proscribe even their possession, at least for "hostile purposes," a significant caveat (Klotz and Sylvester, 2009: 14, 40).

On April 10, 1972, the United States, the United Kingdom, and the Soviet Union signed the Convention on the Prohibition of the Development, Production, and Stockpiling of Bacteriological (Biological) and Toxin Weapons and on their Destruction. The Soviets had initially demanded a convention that covered both chemical and biological weapons before reversing course and joining what appeared to be a unilateral abandonment of an entire class of weapons. Under the terms of the convention, signatory states pledge "never under any circumstance to develop, produce, stockpile or otherwise acquire or retain":

1. Microbial or other biological agents, or toxins, whatever their origin or method of production, of types and in quantities that have no justification for prophylactic, protective, or other peaceful purposes;
2. Weapons equipment or means of delivery designed to use such agents or toxins for hostile purposes or in armed conflict.

However, neither of the superpowers believed that it had an interest in including mandatory provisions for either verification or compliance (Guillemin, 2005: 13, 127, 129).

Both superpowers, and other state actors as well, were satisfied with what essentially amounted to a declaration of good faith that would enable them to continue their most vital work in military biotech with plausible denial. Even the abandonment of an active offensive biological weapons program did not forestall research and development into a new generation of bioweapons under the rubric of defensive research. Arguably, it is even naive to presume that military research programs have spent decades and billions of dollars since the BWC was enacted examining and refining select agents—as the Amerithrax case demonstrated—purely for purposes of defensive research. As Smith (2011) notes, during this time period, the military had largely left biodefense to the public health sector. Ultimately, while nearly every government in the world would accede to the BWC, its primary accomplishment would be to serve as a basis for attempting to reify the norm against BW and to serve as a platform for future collective action (Committee, 2006: 8).

Beyond the Cold War

The BWC demonstrated its limitations early on, when US Senate hearings conducted in 1975 revealed that the CIA had retained various pathogenic stocks, including anthrax and smallpox, as well as deadly shellfish toxins (Guillemin, 2005: 130). Despite the lack of such a mechanism in the BWC, both superpowers demanded inspections of each other's suspected BW facilities, which necessitated both acquiescence and creative explanations for research that violated the spirit of the agreement (Alibek, 1999: 145, 230–245). Given all of these factors, it is unclear whether the BWC regime could have survived a continuation of the Cold War.

> Despite [its] serious verification challenges, the perceived weakness of the Biological Weapons Convention prompted many countries in the early 1990s to call for the negotiation of a legally binding verification regime to supplement the convention. . . . International negotiations began in Geneva in 1995, but major disagreements soon emerged. . . . In mid-2001, after more than six years of talks . . . the United States withdrew its support for the draft Biological Weapons Convention Protocol, prompting widespread international criticism. (Commission on the Prevention of WMD Proliferation and Terrorism [hereafter Commission, 2008]: 35–36)

The George W. Bush administration argued that the verification proposals were incompatible with the security interests of the United States, while permitting greater opportunities for malfeasance by other states that were actively pursuing BW programs. Additionally, "it was widely recognized that none of the proposed measures would have stopped, for example, the anthrax attacks in the United States or any act of biological terrorism by a non-state actor" (Roberts, 2003: 5). Critics contended that such intransigence would cost the goodwill of Russia, China, and other states with defense programs and reduce their incentives to restrain researchers from seeking employment with rogue states and terrorists (Klotz and Sylvester, 2009: 92).

Despite these divisions,

In 2002, at the convention's fifth review conference, the member states agreed to suspend the protocol negotiations indefinitely. Instead, they adopted a U.S. proposal to hold a series of annual expert and political meetings between the review conferences held every five years. Launched in 2003, these annual meetings have focused on the prevention of bioterrorism by addressing such topics as domestic legislation implementing the BWC, pathogen and laboratory security, infectious disease detection and response, scientific codes of conduct, and investigations of alleged use of biological weapons. (Commission, 2008: 35–36)

The BWC, then, codified long-held norms against bioweapons and provided a regime under which targeted cooperation efforts occur (Ball et al., in Maurer, 2009: 494). In the twenty-first century, it remains a more restrictive, but perhaps no more effective, version of the Geneva Protocol.[20] Despite the growth of military biotech programs, no compliance mechanisms or formal organizations exist for monitoring bioweapon research and stockpiles, despite the fact that they are in place for the other types of WMD, nuclear and chemical arms.

Some proponents of novel biotechnological weapons have actually justified their research as upholding the normative aims of the BWC. One Chinese military medical scientist, arguing while advocating for new avenues of biotechnology, perhaps unconsciously echoed pro-BW arguments of a century earlier:

Precision injury and ultramicro damage are two wounding methods of modern biotechnologies based on genomics and proteomics. They are completely different from the traditional wars that damage tissues and organs directly since they target the primary structure of gene or protein. . . . The significance of distinguishing the modern military application of biotechnology from the traditional bioweapons is to promote a healthy development of modern biotechnology, abide by the Biological and Toxin Weapons Convention more effectively, and strike a blow on the traditional bioweapons, therefore welcoming new military progresses and reforms, and changing the notions and civilization level of war. (Guo, 2006: 1152–1154)

PROLIFERATION

The primary interest of the international community pertaining to biotechnology since the Cold War has been preventing biological samples, processing equipment, and technical expertise from being disseminated to other states and to non-state actors such as criminal and terrorist networks. Although the BWC lacks enforcement mechanisms, there have been a number of distinct bilateral and multilateral efforts to reduce the potential supply of bioweapons to rogue entities. It is less clear, however, how much demand exists for BW programs.

One assumption is that the sheer scope of its defense spending has produced a US conventional military so advanced that the only plausible way to attempt to check it is through asymmetric means. This might be through insurgent hit-and-run tactics, but also through attempted deterrence by the threat of chemical, biological, radiological, or nuclear (CBRN) weapons.[21] As a former Indian military chief of staff explained, those planning to engage the United States militarily "should avoid doing so until and unless they possess nuclear weapons" (Preston, 2009: 4).

However, because of the difficulty in developing nuclear weapons and the potentially easy acquisition of naturally occurring pathogens, biological weapons provide an ideal alternative. In many cases, CBRN arsenals are the quickest way that state and non-state actors can legitimize their authority among constituents. It is little wonder that biological weapons are often referred to as the "poor man's nuclear bomb." The observation that biological agents are attractive as cheap WMD highlights their appeal to both hegemonic states and to those who are threatened by hegemony. Wright (2002: 45) argues that "hegemonic realism tends to overshadow moral and legal considerations when it comes to WMD." And in the cases of rogue states and potential competitors for hegemony, the United States is the standard of military force against which these potential proliferators measure their force inadequacies.

Loose Bugs

Recognition of this potential extends back to the Cold War 1960s and played a role in the willingness of the Nixon administration to reprioritize the importance of the strategic nuclear deterrent and abandon bioweapons unilaterally. In the words of Matthew Meselson, the geneticist who began this process in his work for the Arms Control and Disarmament Agency, "Don't we want to make war so expensive that no one can afford it but us?" (Klotz and Sylvester, 2009: 42). The centrality of this concern was evident in the formulation of the "Axis of Evil" by President George W. Bush as rogue states with their own CBRN programs who might potentially serve as proliferation conduits to non-state actors as well (Bush, 2002).

While sanctions have been enacted against rogue states, other methods for curtailing the expansion or diffusion of BW programs are applied toward the private sector. The use of chemical weapons during the Iran-Iraq War spurred the creation of the Australia Group, an international organization whose purpose is the harmonization of export control licensing measures among states with advanced biotech sectors to prevent the dissemination of dual-use technology. Initially a fifteen-nation group, which subsequently expanded to thirty-nine plus the European Union, it succeeded in this capacity in blocking Iraq from obtaining bulk fermenters (Ball et al., in Maurer, 2009: 501–502).

In 2002, the United States unilaterally imposed economic sanctions on several Chinese companies and individual Chinese and Indian citizens for "selling destabilizing arms and germ-weapons materials to Iran on three separate occasions between September 2000 and October 2001" (NTI, 2003). However, the United States itself is conceivably a source of significant biotech dispersal and proliferation. With the formal end of its offensive BW program in 1969, approximately 2,200 American research scientists were left unemployed (Alibek, 1999: 234).

Many of them doubtless found work in the budding private biotechnology sector, which began to expand with the invention of genetic engineering technology in the early 1970s. Today, biotech is a sprawling industry employing thousands of Americans who could potentially be targeted for recruitment by foreign or non-state actors. "Oversight and monitoring of the United States' roughly 1,400 BSL-3 facilities [which handle deadly but treatable pathogens] is currently spread across multiple, poorly coordinated agencies" (Ball et al., in Maurer, 2009: 507).

The threat of "loose nukes," invoked in several twenty-first-century presidential campaigns by both major party candidates as the greatest security threat to the United States, has vastly overshadowed in the public consciousness the potentially equally devastating threat of "loose bugs." Engineered pathogen strains were reportedly stolen during the 1990s in Georgia and Kazakhstan. At the most notorious program site, the BW testing facility on Rebirth Island, pathogens intended for destruction survived and remained potent in the soil well beyond the collapse of the Soviet Union, in part because the cleanup crews cut corners to save costs. The threat that they might be stolen or picked up and transmitted by animals remained a serious possibility, particularly after the Aral Sea shrank so significantly (due to its water being siphoned off for Central Asian irrigation projects) that Rebirth ceased to be an island in 2001 (Preston, 2009: 194). By 2008, there was no longer any body of water in the area, and the former island redoubt melted into the borderlands of Kazakhstan and Uzbekistan.[22]

According to Alibek, the opposite happened to Soviet bioweaponeers: they were suddenly conspicuously Kazakh, Ukrainian, or some other nationality, and many chose to go to their new homelands where they received offers to recreate their work. Alibek stated that he received offers of employment not only from his native Kazakhstan, but also from countries including South Korea, Israel, and France. Other former colleagues reportedly went to Iraq, North Korea, Iran, India, and China (Alibek, 1999: 241, 270–273, 276, 277). Thus the United States claimed that Iranian recruiters offered a salary of $5,000 a month to former Soviet researchers who now made less than $25 a month in Russia, and that similar offers came from India, China, and North

Korea (Preston, 2009: 191). However, Ben Ouagrham-Gormley (2014) notes that, a quarter century after the dissolution of the Soviet Union, there is no evidence of Soviet WMD experts going to work for any rogue state or terror group.

Other colleagues remained in Russia and worked for biotech firms in the new private sector. One such company published advertisements selling genetic material for use in experiments in increasing tularemia virulence, ostensibly for vaccine production (Alibek, 1999: 272–273).

Meanwhile, the United States moved relatively quickly and assertively to prevent the leakage from the former Soviet Union of material and expertise that could be used in developing CBRN weapons. For example, the Department of State established an initiative designed to provide Soviet bioweaponeers with job opportunities in the United States and medical care for relatives who remained in Russia (US House of Representatives, 2005: 15).

In particular, the Nunn-Lugar Cooperative Threat Reduction (CTR) Initiative attempted during the first twenty years of the post–Cold War period to secure or destroy nuclear and biological weapons and safeguard the research scientists who could produce more of them in the future for another employer. Although there was ultimately no evidence that any Biopreparat staff relocated to rogue states—or that their technical expertise could have been employed successfully under other conditions—the United States ultimately spent more than $600 million on commercial research to employ former Soviet bioweaponeers (Ben Ouagrham-Gormley, 2014: 96; Vogel, 2012: 107–109). Chapter 4 examines in greater detail the threat of leaked biological material from the biotech sector.

The Legacy of State BW Programs

Whether or not due to the effectiveness of global governance accords like the BWC and CTR, biotech proliferation from state programs has not been evident. But the continuing research on BW from the late Cold War through the present day demonstrates that the same security dilemma perceptions that drove interwar BW development remain potent today. Most of the countries with BW programs had otherwise formidable conventional forces, and some had nuclear arsenals. Also, the strategic value of BW is questionable at best because there is no documented instance of pathogenic weapons being used to deter a more powerful attacker. Saddam Hussein's Iraq perhaps briefly daunted adversaries prior to the Gulf War, but either refrained from deploying BW when confronted with the threat of nuclear or massive conventional retaliation (Koblentz, 2009: 46) or never had a coherent doctrine or strategy for their use (Sagan, 2013). Thus it would appear that rogue states gained no value from their BW programs, while major powers gained only the fear of falling behind in arms races for weapons that they never intended to use.

Maurer (in Maurer, 2009: 107) notes that between 1950 and 1970, the United States spent $700 million on its BW program, and only had eight antipersonnel weapons and five anticrop agents to show for it. Martin Furmanski of the Center for Arms Control and Non-Proliferation contends that the efforts of the Japanese, British, and Americans managed to confirm that anthrax was the only agent hardy enough to be useful against population centers (Klotz and Sylvester, 2009: 73). And Guillemin (2005: viii) argues that state BW programs were ultimately a "failed military innovation."

However, the industrial-scale research and development programs that produced offensive bacteriological, viral, and toxic arsenals in the last century had lasting consequences, not only in shaping contemporary defense and counterterrorism planning, but also in altering the qualities of entire species of organisms. The previously described practice of increasing the virulence of pathogens by passing them through the immune systems of a series of hosts is one obvious example. During World War II, the British increased the virulence of *B. anthracis* tenfold by passing it through lab monkeys. The Americans succeeded in making the bacteria three hundred times more lethal in the course of their Cold War program, and the Soviets also increased the pathogenicity of previously weaponized stocks (Guillemin, 2005: 107). The *B. anthracis* weaponized during World War II was already an "unusually hardy" strain (Maurer, in Maurer, 2009: 74), as was the Ames strain that would later be grown in quantity and leaked via the Amerithrax attacks. The Soviets bred large quantities of rare strains of smallpox. State bacteriological warfare programs therefore transformed already dangerous microorganisms into far deadlier plagues.

The other significant consequence of the arms race that began with the bacteriological security dilemma of the 1920s was the growth of state infrastructures in unconventional weapons research. For the superpowers, BW programs meant multiple massive facilities, thousands of personnel, and R&D budgets of hundreds of millions of dollars. Without these expenditures, no substantial progress would have been made in developing either bioweapons or biodefenses. The Iraqi BW program provides evidence that committed rogue states without an advanced scientific-industrial infrastructure behind them have not been able to develop game-changing innovations or even to keep pace with the major powers (Maurer, in Maurer, 2009: 74, 81, 107).

In the twentieth century, and in preceding historical periods as well, the exploitation of biotechnologies for warfare was driven by great powers with the resources to harness them far more than it was by weaker adversaries or non-state foes seeking any asymmetric advantage. In both world wars, the interwar years, and the Cold War, states with development capacity explored the potential for BW and then constructed them, out of fear of the consequence of failing to do so when intelligence reported that other actors were also creating them, not because of any inherent strategic value. BW programs therefore followed the classical patterns of arms races, with the advantages going to hegemons, suggesting that the same dynamic would occur with the militarization of other biotechnologies as well.

Ultimately, the biological warfare programs of the twentieth century were extensions of the military adaptations of biotechnology that had preceded them for centuries, but differing in scale and strategic intent rather than being tactical deployments in siege craft. In import, however, these BW programs appear on the surface to have been the strategic dead end ascribed to them by their critics.

But the Cold War BW programs, particularly in the period after the BWC, began to incorporate new advances in biotechnology that went far beyond harvesting germs for slurries. In the early 1970s, just as the international community of states was formally renouncing biological weapons, the international community of scientists was unlocking the secrets of manipulating the genetics of organisms. It did not take long for the new biotechnologies to be incorporated into superpower research programs, which developed entirely new classes of weapons around them, as well as new conceptions of how they might reshape both military and economic power.

NOTES

1. Thomas (2009: 7–8) contends that chemical weapons should not be counted as plausible weapons of mass destruction because they require large quantities and are difficult to use in the field, particularly as compared to biological agents that can be released surreptitiously with deadly results.

2. Despite its toxicity, botulinum has commercial applications as well, being sold in extremely diluted form as the cosmetic enhancement Botox. A factory using bulk fermenters to grow large quantities of the *C. botulinum* bacteria might therefore not have any nefarious purposes related to bioweapons. The dual potential for biowarfare and commercial activity inherent in much of biotechnology is a significant factor in the difficulty in regulating control of illicit material.

3. Lynn Klotz (comments, January 2011) notes that "anthrax, a preferred biological weapon, is not contagious. The fact that it is not contagious is considered an important bioweapon property, as it reduces the likelihood that an attacker will be infected."

4. "Biodefense limits the damage caused by biological weapons" (Smith, 2014: 4), whereas "defense" is interchangeable with military.

5. A similar occurrence afflicted Xenophon's soldiers for several days in the same area centuries before, but they had recovered before the pursuing Persian army reached them.

6. Infected rodents remain endemic in modern Mongolia, with the popular pastime of hunting marmots with trained birds of prey proscribed during summer when contagion is most likely.

7. Another outbreak of plague spread across the Pacific Rim by global commerce in 1894, killing approximately twelve million (Alibek, 1999: 164).

8. Indigenous peoples in South America used the smoke from burning chili peppers as an irritant in both local conflicts and against the Portuguese (CBWInfo, 2005).

9. The Al Aqsa Martyrs' Brigade allegedly packed HIV-contaminated blood into explosives used during the Second Palestinian Intifada (Ackerman and Asal, in Clunan, Lavoy, and Martin, 2008: 187).

10. Japan never adopted the protocol, and the United States did not do so until 1975.

11. American troops also brought hantavirus to the United States during the Korean War. Troop movements are also blamed for the largest smallpox outbreak recorded, with the contagion carried by soldiers demobilized at the end of the Franco-Prussian War in 1871, and for the spread of HIV in Congo, at least partly deliberately, by the army of Zimbabwe (Enemark, 2007: 4–5).

12. Germany's fascist ally Hungary had its own BW program before and during World War II (Dando, 2006: 20).

13. Adolf Hitler had experienced a chemical weapon attack in the trenches of World War I that left him incapacitated and hospitalized. Despite his ardor for using gas against civilians, he ordered that the Third Reich avoid its use on the battlefield.

14. In the midst of the 2001 Amerithrax incident, I often wondered how there could possibly be comparable prior cases of inhalation anthrax for the various agencies to use as treatment guidelines. While some data came from a few cases of "wool sorter's disease" infections, at least some of the expertise had been supplied by Unit 731.

15. Mark Korbitz (2010) notes that "incapacitating agents are particularly useful on the battlefield given the fact that incapacitated soldiers require significant resources and the attention of caregivers, whereas those killed in action do not."

16. Only the strongest pathogens survived the hosts' immune system responses, so by continually passing infected samples between hosts, weaker microorganisms were bred out, leaving the strain more resistant.

17. See the 1989 comic book series *Nth Man* for a fictional account of a bioweapon missile exchange after superpower nuclear arsenals become inert.

18. Recent estimates put the number at approximately seventy confirmed deaths (O'Sullivan in Viotti, Opheim, and Bowen, 2008: 160).

19. Ben Ouagrham-Gormley (2014: 96) states that the figure of sixty thousand employees included all staff, such as janitors; only 10 percent were BW engineers.

20. Stern (2002–2003: 122) argues that the dual-use nature of biotech means that even the tightest regime can only impede BW production, not prevent it.

21. Unconventional armaments were frequently known by the shorthand of *NBC* (nuclear, biological, and chemical) in the 1990s before being replaced by the catchall term *WMD* after 9/11 and during the debate over the status of the Iraqi programs. With emerging concerns over the possibility that radioactive material could be used to inflict extensive localized casualties, radiological weapons were added to the category as well.

22. In 2002, Kazakhstan declared that all testing sites had been decontaminated (NTI, 2002).

Chapter Two

Securitizing Genetic Engineering

The germ warfare arsenals of the twentieth century had little impact on international balances of power because of decisions to pursue more immediate and predictable conventional and nuclear arms. Still, they laid the foundation for new programs for the twenty-first century that could easily dwarf the destructive power of the largest BW usage in history, the low-tech deployment of bacteria that killed tens of thousands in Japanese-occupied China.

But while the science available in the mid-twentieth century involved farming bacteria and viruses in industrial equipment, advances over the past fifty years have opened a new range of biotechnologies. As with airplanes and nuclear physics, it took little time before the most powerful nation-states invested billions of dollars in exploring the security dimensions of these developments even when military uses were far from their original intent.

Advances in genetics have already made engineered pathogens a far greater threat, perhaps to the survival of the human race, than their naturally occurring antecedents, but traditional approaches to BW remain unlikely to become a determinant of international power. Instead, the new biotechnologies offer other advances affording military and economic dominance through readily available food and energy, superior health and longevity, and weapons that can disrupt basic bodily functions without relying on disease transmission. In the twenty-first century, modern great powers with the capacity to field them will enjoy material advantages from biotechnology far more useful than the threat of germ warfare, and smaller rivals will be unable to match production that involves far more than viruses grown in repurposed bulk fermenters and chicken eggs.

Technologies like genetic engineering, and more recently molecular biology, are probably most familiar to students of international security from science-fiction tales (Klotz and Sylvester, 2009: 2). But there is no need to

45

rely on imagination to envision these biotechnologies; they already exist. And as the equipment for the research and manufacture of medicines was adapted for the industrial-scale state bacteriological warfare programs of the last century, many of the developments that can potentially create a new category of genetic bioweapons are also inextricably linked to the health care, agricultural, and research sectors.

Thus, while some of the developments in genetic science are still on the drawing board, there have already been arguments that particular lines of research must be proscribed because they might produce threats to the very survival of the human race. Buzan, Waever, and de Wilde (1998: 23–30) have described how, particularly in the post–Cold War period, policy issues that are not congruent with traditionally conceived military challenges are "securitized" and reframed as threats to national interests. An issue such as genetic engineering undergoes securitization when it is framed "as an existential threat, requiring emergency measures and justifying actions outside the normal bounds of political procedure." So even technologies that remain hypothetical have attracted the attention of the United States, the European Union, and other parties as ultimate security threats requiring the investment of resources to guard against them—often in the form of the further development of biotechnology.

This chapter examines the progression of the genetic revolution that began in the 1970s and how it shifted the focus of BW programs away from the traditional biological arsenals covered under the BWC to potentially new and exotic biotechnologies of warfare. These medical breakthroughs create ethical dilemmas, including in the field of human enhancement for military purposes.

THE GENE AGE

The biotech revolution began with the first successful experiments in gene manipulation in 1973, in which resistance to penicillin was conferred upon a specimen of *E. coli* bacteria. Thus, "with genetic engineering, traditional biological warfare agents can be made more virulent, resistant to antibiotics and vaccines, and better able to avoid detection systems. In addition, harmless microorganisms can be transformed into deadly ones with novel properties" (Koblentz, 2009: 18–19).

The same technology has also offered the means of defeating natural diseases as well. At the same time, biotechnology also promised the tantalizing prospect of improving upon nature's design by eventually creating enhanced, more durable human beings. This revolution has opened a budding bioethics debate, which raises many key points for consideration regarding military applications of biotechnology.

The technology permitting such developments was pioneered in the 1950s, when biologists discovered the process for locating and identifying chromosomes and genes. During that period, the process of separating chromosomes from living cells for observation (karyotyping) was also developed. This process also ultimately makes it possible to insert the genetic information of one organism into a host cell, the basis of cloning. It also laid the foundation for the new process of creating DNA viruses that enter, and then alter, the genetic makeup of already developed organisms. It was this technology that enabled the Human Genome Project twenty years later. The project, which mapped the over three billion genes in human chromosomes, yielded treatments for cancer and various genetic defects before it was even completed (Rifkin, 1998: 8–9). [1]

However, whether such information is used to promote health or harm depends on the intent of the user, and the project also drew criticism for publishing data that could be used in developing biological attacks, particularly against certain groups with distinctive genetic traits. [2] The ability to decode the genomes has also led to the creation of purely synthetic pathogens, including replicas of "the viruses that cause polio and the previously extinct strain of influenza that killed an estimated 40 million worldwide in 1918–1919" (Commission, 2008: 12). [3]

Sheep, Cows, and Chimeras

The process of synthetic genetics has not been limited to modifying individual organisms. In 1973, Stanford biologists reported taking two unrelated organisms that could not mate in nature, isolating a piece of DNA from each, and then recombining the two pieces of genetic material. The process involved the use of a restriction enzyme and ligases that separate individual genes from the chains that constitute the double helix of DNA. The human or animal cells that are isolated in this fashion are also used in cloning.

A comparable length of DNA is then removed from a bacterium or virus that proliferates rapidly and replaced with new cells from the donor organism. This hybrid, called a vector, is used as a means of transferring foreign DNA into the host. The host absorbs the vector, and the DNA it produces now contains that of the host as well. Hybrid organisms are known as transgenic, or as "chimeras" after the mythical Greek monster with the heads of a lion, a goat, and a serpent (Rifkin, 1998: 11).

The potential of recombinant DNA manufacturing for producing radical transformations of species, not least human beings, was not lost on its practitioners at the time. With the controversy about the risk to health and the environment spreading to the general public, in 1974 geneticists imposed a voluntary moratorium on transgenic experiments until the 1975 Asilomar

Conference at which they established industry safety and ethics guidelines (*The Economist*, April 3, 2010).

Still, as with information technology, developments in biotechnology occurred soon thereafter with breathtaking rapidity. In 1979, the "genomic age" began with the first chemically synthesized gene (Dando, in Danzig and Tucker, 2012: 134). During the 1980s and 1990s, researchers claimed that the volume of biological knowledge doubled every five years, and every two in the field of genetics. From this wealth of information came new approaches to the development of engineered strains of plants and animals. Of these, Dolly, cloned from a sheep mammary cell and the egg cell of an unrelated sheep in 1997, was the most prominent example. But she was by no means the only such experiment. Nor was genetic engineering limited to the agriculture industry, although that proved to be the most profitable sector for its application initially (Rifkin, 1998: 12).

In 1983, researchers injected human growth hormones into mouse embryos. When born, the mice grew twice as rapidly, and ultimately twice as large, as any other mice in the population. More significantly, these mice passed their enhanced traits on to their offspring, demonstrating that genetic engineering can have permanent effects on the gene pool of altered species, a fact that must be considered when contemplating the genetic alteration of human beings. The following year, scientists in England combined embryonic cells from a sheep and a goat, creating a chimera that represented the first blending of two completely unrelated animal species in history (Rifkin, 1998: 13–14).

Clearly, this technique could be applied to humans as well, although public outcry following the announcement of Dolly in 1997 led the US Congress to enact a ban on federal funding for human cloning experiments. Apparently, however, this was a matter of closing the barn doors after the cows had gone: in November 1998, a Massachusetts biotech firm announced that it had created a chimera by injecting human cells into a cow egg in 1996 (Anderson, 2007). Although that embryo was apparently not grown to maturity, research into creating human chimeras continued in various laboratories, with researchers in Shanghai in 2003 injecting human cells into rabbit eggs that they permitted to develop for several days before harvesting them for stem cells, and a team in Minnesota that, at roughly the same time, bred pigs that would produce human blood for surgical transfusions (Mott, 2005). Similarly, other researchers have engineered bacteria to produce precursor chemicals for the production of pharmaceuticals (Rutherford and Maurer, in Maurer, 2009: 133).

By the mid-1980s, the private sector had already begun to develop chimeras with commercial utility in manufacturing as well as agriculture. In 1986, recombinant DNA was used to genetically engineer a gene from fireflies into tobacco plants. The resulting plants have leaves that glow with the same illumination as the parent insects, ushering in the potential for producing chemical illumination without significant pollution or heat emission. Other tobacco plants have been engineered to produce human hemoglobin, thus reducing reliance on donor stocks (Adams, 1998), just as human insulin has been produced in vats by bacteria since the 1980s (Judson, 2010). These plants took their place among various other chimeras with applications in the commercial agricultural sector, including engineered predator insects that protected crops from parasites, and aquaculture fish stocks that provided increased food production capacity (the so-called blue revolution).

Another type of fish was created with the security objective of being used to signal exposure to contaminants in water supplies, including from biological agents. Phosphorescent genes were spliced into common zebra fish, creating a population that now passes the trait of glowing neon red, green, or orange (depending on the particular trademarked line) to all of their descendants, which are sold commercially in pet shops as GloFish (Yorktown Technologies LP, 2010).

Such creatures would doubtless be a mere curiosity, perhaps a footnote to the use of regular animals in security or military operations, like carrier pigeons used for communications as recently as World War II, and marine mammals for underwater explosives detection and anti-diver combat operations today (Munoz, 2011; Reuters, 2012; US Navy, 2003). But the same biotechnology was subsequently applied to human medical testing as well. Green fluorescent protein that originated in jellyfish has been fused with human cells for numerous diagnostic purposes, an achievement that won the Nobel Prize in Chemistry in 2008 (Judson, 2010).

The Proliferation of Genetic Engineering Technologies

During this same period, bacteria were engineered to produce bioleaching enzymes that dissolve undesired minerals and leave relatively pure metal ores. The existence of bacteria that can be used to degrade stone and metal presents obvious potential for offensive operations aimed at competitor resources and economies. Biotechnologists have also developed strains of bacteria that consume and break down petroleum, and these have been used to aid in the cleanup of major oil spills dating back to the 1989 Valdez, Alaska, disaster. Similarly, scientists have developed strains of *E. coli* that consume organic waste and secrete it as fuel-grade ethanol, a development with a potentially significant impact on the international oil trade (Rifkin, 1998: 14–17).

In another vein of research, scientists have used engineered viruses to produce catalysts that mimic the natural process of photosynthesis used by plants to generate energy by splitting water into hydrogen and oxygen (Chandler, 2010). While the quantity of energy generated by this process is minuscule, the development of this biotechnology opens the theoretical possibility of eventually generating large amounts of clean renewable energy, which would similarly impact current fuel-exporting states. Further, in 2007, a former Human Genome Project team announced the transplantation of the entire genome of one species of bacteria into another with the goal of creating organisms that would produce biofuels or consume CO_2 emissions (Preston, 2009: xx).

Bacteria are also being designed for the purpose of biosorption. Naturally occurring biomass such as fungus has been used since the early 1990s to absorb harmful chemicals at industrial accident sites. Subsequently, geneticists developed bacteria that can also absorb heavy metals and radionuclides (Rifkin, 1998: 17–22; Vijayaraghavan and Yun, 2008: 266–291). A related biotechnology with applications both in the field and at home is bioremediation, which employs enzymes or plants to absorb or break down materials. The US Department of Defense has sponsored research into phytoremediation of hazardous materials using plants (Fox, 1997) and called for a renegotiation of the BWC to permit the development of material-degrading microorganisms (Sunshine Project, 2008). Eventually, bioremediating organisms might be used by militaries, or firms in acts of industrial sabotage, to degrade competitor resources such as fuel supplies.

Another lucrative area in genetic engineering is biomedicine, such as the livestock that produce blood and insulin suitable for transfers to humans. A more recent development is the grafting of silicon-based computers directly onto human or animal tissue "by attaching it to engineered silkworm cocoons—this mixture is capable of attaching directly to tissue, with the silk dissolving and leaving the circuit attached directly to tissue. . . . [This] research [is] being used to monitor animal brains for neural activity in epileptic animals, possibly leading to an eventual cure for epilepsy, or even drug delivery or electrical stimulation to damaged nerves within the body" (National Institutes of Health, 2013). In future wars, casualties may quickly receive skin and organ grafts from cells manufactured by engineered organisms as well, and civilian patients are already receiving grafts of human stem cell tissues created by 3-D printing. A more detailed description of military R&D in this field and the likely consequences follows in the next chapter.

Genetic engineering is also being used for the production of synthetic materials in manufacturing. Some bacteria produce minute quantities of durable polymers, and researchers have succeeded in breeding strains of bacteria capable of producing plastics with variant degrees of elasticity. Since the 1990s, biotech firms have spliced genes from polymer-producing bacteria into mustard plants in efforts to grow plastic for commercial use. The US Army is developing bacteria containing the genes of orb-weaving spiders, whose silk is among the strongest fibers known to exist. The bacteria are intended to be grown in sufficient quantities to harvest the silk for use in, among other products, bulletproof vests and parachutes (Rifkin, 1998: 15–16).[4]

Still another key area of biotechnology is enzymology. Enzymes are protein catalysts capable of altering the rate of a chemical reaction, including how membranes transport proteins. Enzymology is used to develop membranes that effectively shield against microscopic quantities of infectious agents (Sylvester and Klotz, 1987: 17, 42, 81). The same biotechnologies that have generated new unconventional security threats have therefore also produced the means for countering some of them.

The Promise and Peril of Synthetic Biology

In 2010, biologist Craig Venter, a pioneer of the Human Genome Project, announced that his team had created the first cell with a synthetic genome, a copy of a bacterium made from off-the-shelf laboratory chemicals and inserted in the empty shell of another type of bacteria. "The result is the first creature since the beginning of creatures that has no ancestor. . . . When the first of these artificial creatures showed that it could reproduce on its own, the age of artificial life began" (*The Economist*, May 22, 2010, "Genesis Redux").[5] By 2013, Venter was developing a new microorganism to be created completely from synthetic design (Coghlan, 2013).

The biotechnology underlying this Frankenstein-like creation is synthetic biology, which will ultimately permit the production of any type of pathogen or microorganism, including those being considered for military or industrial purposes. It will also blur the distinction between living creatures and man-made materials, creating a host of ethical and practical questions at all levels of human activity, not least in defense planning.

> Crucial to [synthetic biology's] realization as an economic sector is the development of standardized and interchangeable biological units known as bio-bricks. . . . In essence, bio-bricks are to biology what the standardization of electronic circuits was to the global information technology industry in the 1970s. To build a global synthetic biology commercial sector requires a literal cataloguing of standard biological units that can be manipulated and fitted to suit each new product/application, while remaining seamless and interchangeable at the most basic level. (Fletcher and Allen, 2007: 1, 14–15)

While geneticists demonstrated in the 1940s that it was possible to alter the heritable traits of living organisms by using radiation to produce mutations, advances in this field were limited by the long process of decoding DNA to determine exactly which genes controlled particular functions. But the development in the 1980s of polymerase chain reaction (PCR), which allows the splitting and copying of genetic material to produce many more samples than would be available naturally, accelerated research tremendously (National Research Council, Committee on Opportunities in Biotechnolo-gy for Future Army Applications [hereafter Committee, 2001]: 11; Specter, 2010).[6]

But more than that, PCR caused "mixtures of uncoupled [DNA] strands to randomly pair with one another rather than with their twins . . . [allowing researchers to] then select the new DNA for specific characteristics, much as dog breeders select for hunters or racers." Geneticist Willem Stemmer, who discovered this technique, was consequently able to "quickly 'direct evolu-tion' to select for an *E. coli* that was 32,000 times as resistant to a given antibiotic as the *E. coli* with which he had started" (Specter, 2010).

With the development of clustered regularly interspaced short palindrom-ic repeats (CRISPR) in the early 2000s, geneticists could now create herit-able germline changes through direct editing, adding or removing specific genes of their choice, although altering human embryos remained a treatment that was not expected to eventuate until the 2020s. Still, modified "Crisprs" immediately offered the potential to fulfill the goals of Biopreparat: one leading researcher reported that she needed to halt a student lab project that introduced human lung cancers to mice through mutation delivered by "Crisprs that could be packed into a virus and inhaled" because a minor coding error could have made the vectors affect humans (Kahn, 2015).

The newfound abilities of geneticists to identify, isolate, and synthesize genes has led to other rapid advances, with biobricks from multiple organ-isms inserted into *E. coli* bacteria to create chimera that grew an antimalarial pharmaceutical and an enhanced, hardier strain of the polio virus during the early 2000s (Specter, 2010). Geneticists have also developed purely synthetic "XNA," alternative polymers that contain genetic information, can be made to reproduce in laboratories, and can evolve through natural selection. They should not be able to interact with naturally occurring DNA and RNA be-cause of their different biochemistries, eliminating the possibility of genetic contamination, and are largely "invulnerable to enzymes, extreme pH values, and other harsh conditions" and can be used as pharmaceutical delivery systems (Yong, 2012). But, because "an XNA organism can potentially harm an RNA/DNA organism (such as humans) if XNAs could somehow get into our own genetic information," such durable agents would obviously have the capacity to deliver biological attacks (Xu, 2013).

The field of synthetic biology is expanding rapidly, and researchers are promoting its principle of modularity, some comparing it to literal building blocks of life, for the potential to crowd-source genetic development. The BioBricks Foundation, a nonprofit organization formed to register and devel-op standard parts for assembling DNA, supports some of this research (Mooallem, 2010). Among the projects is an international biobricks registry that is both a physical repository and an online catalog:

If you want to construct an organism, or engineer it in new ways, you can go to the site as you would one that sells lumber or industrial pipes. The constituent parts of DNA—promoters, ribosome-binding sites, plasmid backbones, and thousands of other components—are catalogued, explained, and discussed. It is a kind of theoretical Wikipedia of future life forms, with the added benefit of actually providing the parts necessary to build them. (Specter, 2010)

In 2010, the National Science Foundation sponsored the creation of a biobrick production facility called Biofab for the purpose of making modules widely available to researchers. A firm called LS9, with financing by oil company Chevron, has already used commercially available biobricks to develop bacteria that convert sugar into "a chemical compound that is almost identical to diesel fuel [that] the company calls a 'renewable petroleum.' Another firm, Amyris Biotechnologies, has similarly tricked out yeast to produce an antimalarial drug." Many of these companies sponsor university and high school student research teams in global competitions, with one team developing a synthetic yeast that lights up in response to electricity that could be used to produce cheap computer displays. At one competition, "the backs of some team shirts overflowed, NASCARlike, with their sponsors' logos, including those of a few multinationals like Monsanto and Merck" (Mooallem, 2010).

By offering a growing array of researchers and institutions the ability to alter entire organisms, and through them species, synthetic biology represents a shift from the earlier biotechnology of genetic engineering, which permitted changing only individual genes. Craig Venter is modifying algae to triple the amount of petroleum that can be derived from them (Coghlan, 2013). Other researchers are exploring the possibility of recreating extinct species such as mammoths or even dinosaurs. DNA from the extinct Tasmanian tiger was made operative again when inserted into a mouse embryo in 2008, "the first time that any material from an extinct creature other than a virus has functioned inside a living organism" (Specter, 2010). Given that geneticists now have recreated extinct pathogens as well, the securitization of synthetic biology is already at hand, with a rising chorus of calls to restrict access to data because unchecked developments in the field hold the potential to create a novel virus that is lethal, transmissible, and incurable (Rutherford and Maurer, in Maurer, 2009: 115).

The Potential of Human Enhancement

But while some critics charge that synthetic biology could conceivably cause the extinction of the human race via an unstoppable engineered viral pandemic, others worry that seemingly benign products of genetic engineering could mean the effective end of the species by transforming it into something else. The ability to modify viruses and clone livestock means that scientists ultimately can, and perhaps already could, change the human body and potential with germline modifications that would alter traits in their descendants as well. Why not, then, relieve human beings of suffering, prolong their life spans, and also give them abilities beyond those they naturally possess to more fully succeed in their lives? What would it mean if the entire human race is enhanced? What would it mean if only a fortunate few were?

After decades of scientific advances, an array of literature examines the potential impact of genetic engineering on questions of social as well as medical ethics. But, somewhat surprisingly, there has been little more consideration by ethicists about the impact of human enhancement on military personnel or their targets in war zones than there has been by political scientists about the potential effects of genetic engineering on international affairs.

Political philosopher Michael Sandel (in Savulescu and Bostrom, 2009: 73–74) examines the question of performance enhancement in sports and asks what precisely we find troubling about the idea of genetically altered athletes performing superhuman feats. He contends that it cannot be attributed to a sense of unfairness, because competitors "naturally have different physical capabilities to begin with." In his discourse on the enhancement of athletes, and also potentially businessmen with augmented linguistic abilities for competitive advantage, Sandel fails to question the ramifications of enhanced soldiers competing on the battlefield, or of entire nations with artificially increased aptitudes and what that might mean for global economic and political power. Obviously different states have varying levels of resources and capabilities to begin with, so should there be anything troubling about using biotechnology to extend those relative advantages either economically or militarily?

The augmentation of soldiers is not a topic explored in the otherwise wide-ranging debates between pro-enhancement transhumanists and anti-enhancement bioconservatives (Bostrom and Savulescu, in Savulescu and Bostrom, 2009: 1). In one prominent expression of bioconservatism, political philosopher Francis Fukuyama chose transhumanism as his selection for an issue of *Foreign Policy* dedicated to "The World's Most Dangerous Idea." Arguing that a raft of existing medical technologies from pain blockers to neonatal screening will likely soon lead to norms of human enhancement that are not questioned until it is too late, Fukuyama argues that "the first victim of transhumanism might be equality." He notes that the implications for uneven distribution of health and ability, and consequently wealth and power, are more troubling at the international scale because of the already wide differential between the most and least developed countries (Fukuyama, 2004: 42–43). Opposition to human enhancement as a potential threat to the survival of humanity has been driven by merging of the arms control and environmental movements with the issue of genetic modification (Juengst, in Savulescu and Bostrom, 2009: 48).

The opposing, transhumanist view is articulated by futurist Ray Kurzweil, who singularly disagrees with Fukuyama about the definition of humanity. He however rejects the term *transhuman* out of a belief that it is perfectly human to improve and transcend existing abilities, and expresses few qualms about incorporating technology directly into the body to do so. Kurzweil argues that the transformative power of technology on human society is a positive one because of enhanced economic productivity and communication abilities. He dismisses Fukuyama's concern about the potential for exacerbating global inequality with the contention that "eventually these technologies reach everyone." But despite his overall optimism, Kurzweil also sees destructive potential in new biotechnologies, describing the possibility of an engineered supervirus as "an existential risk" (Kurzweil, 2005).

David Ewing Duncan (2012) considers the ethics of augmentation when everyone else is doing it. Parents might initially balk at the idea of using technology to increase their children's cognitive performances, but not if it means that they fall behind their augmented classmates. Citizens might hesitate to vote for presidential candidates with neural implants to enhance their reflexes and decision-making capabilities during a crisis. But at some point, the question becomes, "Would you vote for a commander in chief who wasn't equipped with such a device?" Ultimately, and particularly if rival hegemons are dispatching Augments with advanced bioweapons and biomedicines to the battlefield, what country with the capability to do so could justify sending its soldiers into harm's way without the best advantages possible?

In a 2013 report for the Greenwall Foundation, which researches policy questions related to bioethics, Patrick Lin, Maxwell Mehlman, and Keith Abney offer perhaps the most comprehensive examination to date on soldier enhancement through biotechnology. Their definition includes biological or medical interventions in the body designed to improve performance beyond what is necessary to sustain normal health. Their primary concern is the fate of individual soldiers who may be required to accept augmentation under military conditions that do not conform with the informed consent necessary for civilian patients. As much current research into soldier enhancement concerns blocking symptoms of pain and fatigue, there is the potential for harm to enhanced subjects that outweighs the benefits to them. Still, military personnel may ultimately be left with no choice, either through direct orders or by the pressure of not advancing in their careers while enhanced comrades in arms receive assignments and decorations because of their new abilities. The "warrior's code" of soldiering may itself be threatened if augmented warfighters are perceived as something other than truly human by political leaders, by members of society including family—potentially making it more difficult to rally mass public support for war—or even by themselves or other troops, which would threaten force cohesion, or by enemy forces who may treat them with less mercy (Lin, Mehlman, and Abney, 2013: 9–11, 15, 39–49, 79–81).

GENETICALLY ENGINEERED BIOWEAPONS

In the case of human enhancement or any other technology dating back to the BW research of the past century, the intent of the wielder and the threat perceptions of potential targets are the only markers of offensive biotechnologies. For example, the Soviet Biopreparat set out to create chimerical "designer" pathogens not found in nature that would combine the effects of multiple diseases and be resistant to antibiotics, vaccinations, and antivirals. One documented project was the creation of Legionnaire's Disease combined with myelin toxin that would produce fatal symptoms resembling multiple sclerosis, although such investigations would have been for purely defensive purposes under the BWC rubric (Preston, 2009: 184, 295). But this research also ultimately led to human insulin becoming affordable to produce in mass quantity with the discovery that it could be grown in laboratories by transferring its genes to bacteria (Alibek, 1999: 157). Security threats from genetically engineered pathogens and vectors used to overwrite existing genetic codes can originate from state BW and biodefense programs, or from commercial research.

State BW Programs

As with bacteriological weapon production, the Soviets had what was apparently the largest and most advanced program of genetically engineered weapon manufacture. After decades of Soviet dismissal of Mendelian genetics as bourgeois science, an ambitious budding party apparatchik and accomplished biochemist persuaded President Brezhnev in 1971 that they were at risk of falling behind the West in what Soviet analysts were describing as the Military Technical Revolution (known as the Revolution in Military Affairs in the United States and discussed in the next chapter). Yuri A. Ovchinnikov rapidly transformed a moribund BW infrastructure that was being considered for mothballs in the wake of the suspension of the American BW program into a robust platform for the development of new genetic weapons. In 1973, Biopreparat launched a gene-splicing program called Ferment, but referred to in English as Enzyme (Leitenberg and Zilinskas, 2012: 9, 51–59), which would develop genetically altered pathogens that would be deliverable by ICBM. Another program called Metol sought to engineer the bacteria to resist antibiotics (Alibek, 1999: 41, 159).

Still another project called Bonfire used advances in peptide synthesis to create new toxin weapons. Mirroring the dark potential of biomedicine and bioremediation, the biotechnologies were used to manufacture toxins that were difficult to obtain in large quantities, such as from exotic snakes and fungi. Another goal of the research was to create pathogens, including plague, that also produced myelin toxins that would create serious damage to the brain and spinal cord.

Similar work was effective in producing excesses of the bioregulators that control human organ functions, with heavy imbalances of peptides producing heart palpitations and death. As Alibek noted, "a new class of weapons had been found. For the first time, we would be capable of producing weapons based on chemical substances produced naturally in the human body. They could damage the nervous system, alter moods, trigger psychological changes, and even kill." Biopreparat's ready response to the potential discovery of their work by foreign intelligence was that "weapons based on compounds produced in the human body were not prohibited by the BWC" (Alibek, 1999: 154–155).

Indeed, while some scholars include altering bioregulators in the definition of biowarfare, the BWC refers only to pathogens and toxins. Bioregulators, which include serotonin, endorphins, and insulin, affect not only physical bodily functions but also moods and psychological stability. Unlike bacteriological weapons, they do not rely on microorganism reproduction, so the effects would be felt within hours of exposure, sufficient time to affect battlefield conditions (Koblentz, 2009: 9–10).

A 1997 Department of Defense study imagined six potential threats from emergent biotechnology, five of which were modified pathogens and one was "gene therapy used as a weapon." Because of the dual-use potential of bioregulators, in the twenty-first century, "the concept of a 'biothreat agent' will expand beyond the current limited perspective of biothreats as being only 'bugs' (i.e., pathogenic organisms) to include an entirely new category of threats, the biological circuit disruptors" (Committee, 2006: 39, 139).

> One of the more concerning assaults, yet attainable even with today's delivery technology, could arise from the use of targeted delivery systems to insert genes into chromosomal DNA. For example, viral delivery vectors developed for human gene therapy exploit the ability of viruses to build selectively to specific cell types as a way to deliver genes encapsulated inside the viral particle into the target cells. . . . The disruptive effects could be manifest immediately as an acute event or the gene could lie silent within the genome for activation at a later time by a second external trigger. . . . Potential delivery platforms include the use of bacterial plasmids or viral vectors for cloning the genes that encode bioregulators; the use of transgenic insects (i.e., to secrete and inoculate the bioregulators); nano-scale delivery systems (e.g., engineered proteins either within or bound to nanotubes). . . . Moreover, transgenic plants could be put to dual-use as bioregulator production factories. (Committee, 2006: 149, 154)

Current military research into direct-effect weapons is described in the next chapter.

Military applications of chimera were of particular interest to the Soviets, with bioweapons developed from multiple pathogens, including in a project called Hunter that was intended to yield both bacterial and viral infections (Leitenberg and Zilinskas, 2012). Others included the incorporation of liegu fever vectors into wine yeast and the creation of an influenza virus that would also produce a toxin found in cobra venom (*Shenyang Liaoling Ribao*, 1997).

Using plasmids, strands of replicative genetic material, as transmission media, Biopreparat tested such chimera on rabbits, which reportedly both developed plague symptoms and experienced paralytic effects. As the inventors recognized, in the course of an attack, the source of the neurological damage would not be evident without knowledge that the toxin was being produced by the altered plague. Other projects included attempts to insert Ebola genes in smallpox viruses and the Bonfire-like production of beta-endorphins "capable in large amounts of producing psychological and neurological disorders and suppressing certain immunological reactions" (Alibek, 1999: 160–164, 258–263). The latter category included tularemia that also produced painkilling endorphins that would render subjects catatonic before they could realize their distress (US House of Representatives, 2005: 8–9).

Alibek (1999: 258–263) would later argue that there was evidence that Russia reconstituted its genetically engineered BW program in the late 1990s, and also that the technology was proliferating, as his former colleagues published their research involving chimera with no evident commercial applications. However, other testimony and records from Biopreparat personnel indicate that the Soviets ultimately made only minor progress developing viral chimera and did not actually modify any select agents such as Marburg. "Soviet bioweapons scientist Igor Domaradskij writes in his memoir that the Soviet bioweapons program essentially failed to produce the new generation of bioweapons it set out to develop in the 1970s despite a prodigious investment in human, material, and financial resources." While the results may simply have been technologically unfeasible for the Soviets, researchers were prevented by political and security constraints from acknowledging the problems that they encountered in their labs (Ben Ouagrham-Gormley, 2014: 37, 96). Conceivably, high-ranking officials such as Alibek received overly optimistic reports about the outcomes of Biopreparat R&D.

Perhaps Biopreparat's most dangerous legacy is the inspiration its work on genetically engineered BW provides to other actors attempting to supersede it. In China, *The People's Military Surgeon* publication of the People's Liberation Army has regularly run articles on the development of the military biotech field. Some reports have stressed that "the focus of present and future research will be placed on the killing and wounding effects of various types of new weapons and patterns of treatment" and that "they will be put into use in the next 3–15 years" (Guo, 1996).

An article produced by the "No. 1 Military Medical Sciences Laboratory" notes that with existing production and refining equipment, the manufacture of biological weapons can be moved from large-scale production facilities to small laboratories, making detection far more difficult. The article also notes that the development of biotoxins is much easier than the development of preventative measures, and that producing bioweapons therefore is an ideal cost-imposing strategy on competitors (Wang, 1998). Other possibilities for thwarting biodefenses include engineering bacteria designed to make it difficult for the immune system to affix antibodies, or using the science of bioinformatics to develop pathogens that cause the symptoms of familiar BW like anthrax but that are genetically dissimilar and can therefore escape detection or defeat stockpiled treatments (US House of Representatives, 2005: 22, 34, 39).

Further, China's military laboratory reports that "the pharmaceutics industry is now carrying out conscientious research on methods to make drugs stable, so as to be able to use aerosol to dispense toxins and immunological regulators." The shape of Beijing's long-term biotech strategy is discernible through the inclusion in the same article of a reference to the BSE protein that causes "mad cow disease." The report notes that although the infectious potential of BSE is very low and its latency period quite long, even a few cases were sufficient to decimate the British beef industry (Wang, 1998). The implication is quite clear: biotechnology can be used as a weapon of industrial sabotage, and the actual efficacy of the agent used is not necessarily as important as its value as a terror weapon, the same argument made by Western military researchers a century earlier.

China has also studied the feasibility of breeding insects engineered to produce a variety of viruses (Wang, 1998). Previously, Canada's Cold War BW program had "focused on insect vectors to deliver diseases" (Smithson, 2011: 231). Insects could be an ideal delivery system for delivering incapacitating agents prior to an attack, particularly now that they can be directed remotely through implants (see the next chapter). "Super pests" could be engineered to withstand pesticides and be directed against enemy agriculture. A more radical idea is to use these vectors to deploy existing microbe technology to create organisms that will attack the metal, rubber, and plastic of production facilities or command, control, and intelligence systems (Adams, 1998). Petro, Plasse, and McNulty (2003: 163–164) note that "transgenic insects, such as bees, wasps, or mosquitoes, could be developed to produce and deliver protein-based biological warfare agents . . . [and] because many bioregulators and toxins are thought to be effective at exceedingly low doses, an individual may succumb to infection after having been bitten by a few transgenic mosquitoes."

Oversight of Genetic Weapons Development

Between the three hundred different genetic engineering projects operated by Biopreparat and subsequent advances by state military programs and by private-sector genetic engineers and synthetic biologists, a number of alarming scenarios for twenty-first-century biological attacks are imaginable. And while it took Soviet bioengineers a full decade of research before they were able to engineer pathogens in the 1980s, technological advances and the proliferation of knowledge and expertise have made it far easier for other actors to follow in their footsteps (US House of Representatives, 2005: 6–7, 22).

Because states have cataloged the characteristics of the naturally occurring pathogens that would be likely to be used in BW programs, they can prepare defenses against even bioweapons bred for increased pathogenicity. However, this would not necessarily be true of chimeras, and some estimates are that non-state actors could engineer drug-resistant or more virulent pathogens with three to four years of research. "More advanced weapons would require terrorists to manipulate behaviors controlled by multiple genes, extract substantial amounts of DNA from two genomes to create a fundamentally new, third organism or program computer-like behaviors that focused pathogens on narrowly defined targets" (Rutherford and Maurer, in Maurer, 2009: 131, 133, 137).

An example would be the plasmids that typically carry the genes that cause virulence in BW bacteria, such as those that cause plague and anthrax. In some cases, plasmids produce deadly toxins, but only when exposed to particular biochemicals, with one example being a strain of *E. coli* that will produce a toxin that causes hemorrhagic enteritis, but only in intestinal tracts. While humans have struggled to create binary chemical weapons, nature has already provided deadly biological binary weapons that "make it feasible to grow up to kilogram quantities of reagents without posing an undue risk to its manufacturer" (Block, in Drell, Sofaer, and Wilson, 1999: 56).

Other biotechnologies could also multiply the effectiveness of initially rudimentary bioweapons, with enzymes enhancing the ability of vectors to penetrate skin, engineering to make detection more difficult by changing particular genes, changes in incubation periods to confound quarantines, or reductions in vulnerability to established treatments, thus overwhelming target governments' stockpiles of vaccines and antibiotics. Non-state actors can purchase transgenic organisms or purchase the services of biobrick synthesis companies, which, despite their security screening efforts, might be deceived by novel attempts to create bioweapons. Unintended experimental results have provided alarming evidence that it is easy, without even trying to do so, to increase the pathogenicity of a virus to the point where it can overwhelm all available defenses (Maurer, in Maurer, 2009: 78, 108).

One effort in Australia in the 1990s saw experiments that were intended to combat a periodic infestation of feral mice by developing an engineered mousepox vector as a vehicle to make mice sterile and then transmit the virus to others. But they had quite a different outcome. Adding the biobrick to the virus stimulated the virus enough to make it deadly even to mice that had already been immunized. The researchers had not been expecting this sort of bioweapon and quickly realized that such an enhancement could be accomplished with the smallpox virus as well. Given that "300 million people died from smallpox in the twentieth century alone—roughly three times the number who died in armed conflicts"—the potential costs of so-called "black biology" involving pathogen virulence, even if not formally part of a BW program, could be catastrophic (Cohen, 2002).

Maxygen, the company created by Willem Stemmer, the geneticist who discovered how to increase pathogenic virulence by using PCR to produce rapid evolution,

> illustrates the hairline borders between bio-medicine, bio-weapons, and bio-defense. . . . Maxygen has built its business around the search for new drugs, vaccines, disease-resistant plants, and industrial biochemicals. In addition to helping other companies speed the creation of better drugs to treat allergies, multiple sclerosis, and psychiatric diseases, it is working to develop its own therapies for auto-immune diseases and cancer [and HIV]. And the company has several contracts with the U.S. Defense Advanced Research Projects Agency (DARPA).[7] (Cohen, 2002)

Such contracts are one vital feature of the new biodefense industry, which had previously been the provenance of state laboratories and now provides not only new points of research entry but also exit to proliferation and misappropriation through leaks via the private sector. And even synthetic biology companies with no evident ties to military research might provide the materials needed to manufacture select agents to rogue groups or individuals. Because the new biotechnologies have the same dual-use potential as those of the previous century, both governmental and industry sources have issued calls for regulation of commercial research in the interest of security.

> [DNA synthesis is now an international industry that] currently fills thousands of orders for whole genes and millions of orders for short oligonucleotides each day. In some instances, these orders comprise DNA sequences derived from pathogenic organisms, and in rare cases from pathogens with an extremely high mortality rate, such as the Ebola virus. So far, none of these orders has been identified as malevolent. . . .

Nevertheless, it is important for gene synthesis companies to detect such orders for genes of concern effectively and efficiently. To achieve this, almost all gene synthesis houses apply a screening process in which incoming orders are homology matched against a database. However, reports persist that a few gene synthesis companies have yet to embrace this practice and this loophole needs to be closed. (Industry Association Synthetic Biology, 2008: 4–5)

Meanwhile, the United States' Commission on the Prevention of WMD Proliferation and Terrorism (2008: 39) has called for new efforts at global governance beyond the control of traditional biological weapons, with advanced industrial states

that possess advanced capabilities in the life sciences (e.g., Canada, France, Germany, Japan, Switzerland, the United Kingdom, and the United States) and emerging biotech powers (e.g., Brazil, China, India, Malaysia, Singapore, South Africa, South Korea, and Russia) to develop a road map for ensuring that the revolution in biology unfolds safely and securely.

The purpose of such a biotech powers conference should be to identify key principles of biosecurity, to harmonize national regulatory frameworks for dangerous pathogens and dual-use research of concern, and to promote international biosecurity cooperation. . . . Furthermore, the conference would consider bottom-up approaches for raising the awareness of life scientists in academic institutions and commercial enterprises about the security dimensions of their work, with a view to creating a transnational "culture of security awareness."

ASSESSING INTERNATIONAL SECURITY AFTER THE BIOTECH REVOLUTION

In the 2010s, the international regime for controlling biowarfare remained grounded in assumptions of naturally occurring bacteria churned out in bulk fermenters, but a new wave of biotechnology as old as the regime itself had made other forms of BW more threatening. Chimerical microorganisms, superevolved viruses, and direct-effect weapons targeting human bioregulators were the emerging threats against which new defenses needed to be established. And while their development may have been pioneered by the Soviet Union, the fear was that they were theoretically within the reach of virtually any entity able to purchase biobricks from commercial registries.

Particularly in the decade following the 9/11 and Amerithrax attacks, for many, the concern was that exotic and lethal new bioweapons would be employed by non-state actors such as terrorists, particularly if they were religiously motivated or otherwise martyrdom seeking and therefore seemingly undeterrable.[8] However, as Gregory Koblentz has noted, there are two shortcomings to such assessments of the threat of engineered microorganisms. One is that terrorists have no motive to expend the resources to develop engineered select agents because those that occur naturally are already lethal and terrifying enough. The other is that these analyses make the mistake of conflating the ability of non-state actors, even with technological proliferation, to match the research capacity of states (Koblentz, 2009: 214–215). Still, these criticisms overlook the capacities of another type of non-state actor: research labs, whether based in corporations or universities, that have been hired to perform sabotage against agricultural, manufacturing, or population centers, or that have been contracted by states to develop military technology.

Despite proliferation and the democratization of access to biotechnologies, the evidence indicates that the main research and stockpiling of bioweapons and biodefenses will continue to be state military programs. Hegemons that rely on military strength, as well as a handful of rogue states that fear attacks by conventionally superior adversaries, have been the primary producers and consumers of CBRN weapons, and they are likely to continue to be so with military applications of genetic engineering as well. Indeed, military researchers are already reaping the harvests of the genetic age and forging ahead with a wide array of new biotechnological enhancements of military power, with implications for both the stability of international relations and the future of the human race itself.

NOTES

1. The project ran from 1990 until 2000, when a working model of the roughly twenty-five thousand genes in the human genome was published by one of the project research teams. The US government sponsored the research, but parallel and follow-on efforts were conducted by private biotech firms.

2. It is not clear that such weapons are feasible, despite the efforts of South Africa to create them, described in the previous chapter, and rumors of similar Iraqi and Israeli efforts, described in the next chapter. However, certain disorders, such as sickle-cell anemia and Tay-Sachs, overwhelmingly afflict only particular ethnic groups, and some medications, such as primaquine, likewise only cause severe side effects in patients with the genetic markers prevalent within particular ethnic groups.

3. Complete genomes for plague bacterium were decoded in 2001 (Preston, 2009: 296) and, in just six weeks of research, for SARS in 2003 (Koblentz, 2009: 231).

4. The Canadian firm Nexia has manufactured spider silk using recombinant proteins from dragline spiders and bred goats to manufacture larger quantities of this protein in their milk (Egudo, 2004: 7).

5. Lynn Klotz (in January 2011 comments) notes that *The Economist* mischaracterized Venter's accomplishment, as the engineered bacterium had an ancestor in the closely related bacterium from which its genetic material was taken.

6. "Commonly referred to as a molecular photocopier, PCR works by heating DNA until the strands of the double helix uncoil. Chemicals are then added and the temperature is lowered, allowing each strand to replicate itself and form a new double helix. Repeated many times over, the process can turn minute amounts of DNA into large quantities in a matter of hours" (Cohen, 2002).

7. Some of DARPA's numerous biotech projects are described at length in the next chapter. Among the contracts with Maxygen was a broad-spectrum aerosol vaccine (Maxygen, 1999), a counter-BW measure first proposed by the French in the 1920s.

8. The question of whether there is such a thing as a truly undeterrable terrorist is not settled, as individuals willing to sacrifice their own lives for a cause might balk at putting their families in the path of retaliation, or even their entire society by nuclear or massive biological retaliation. Given that advances in genetic technology have also made it easier to trace the origins of select agents, it seems unlikely that anonymity and a lack of fixed location as a target to attack would prevent retribution forever. (See chapter 4 for how the Amerithrax culprit was traced.) Still, it is plausible that genetic engineers could obscure or erase the types of genetic markers used in forensic microbiology (Korbitz, 2010).

Chapter Three

Twenty-First-Century Military Biotechnology

Imagine soldiers having no physical limitations . . . water and power being available whenever and wherever they are needed . . . mechanical systems as autonomous and adaptable as living things. What if, instead of acting on thoughts, we had thoughts that could act? Indeed, imagine if soldiers could communicate by thought alone . . . communications so secure there is zero probability of intercept. Imagine the threat of biological attack being inconsequential. And contemplate, for a moment, a world in which learning is as easy as eating, and the replacement of damaged body parts as convenient as a fast food drive-thru. As impossible as these visions sound or as difficult you might think the task would be, these visions are the everyday work of the Defense Sciences Office. . . .

Enhanced human performance . . . is born from the realization that with the emphasis on technology in the battle space the human is rapidly becoming "the weakest link." Soldiers having no physical, physiological, or cognitive limitations will be key to survival and operational dominance in the future. . . .

The exoskeleton initiative will provide mechanical augmentation extending individual performance. Metabolically dominant warfighters of the future will be able to keep their cognitive abilities intact, while not sleeping for weeks. They will be able to endure constant, extreme exertion and take it in stride. Success in metabolic engineering will be visible, because I will be the first volunteer to be transformed.

—Michael Goldblatt, Director, Defense Sciences Office, DARPA (2002)

The development of genetic engineering in the 1970s and the successful achievement of creating entirely artificial life forms at the beginning of the 2010s have expanded the scope of biotechnologies dramatically. And the effects—such as a greatly reduced proportion of American deaths from combat wounds in the Iraq War—will significantly impact both the propensity of hegemonic states to use force and the ability of less powerful actors to offer resistance.

Military planners in advanced industrial states continually issue optimistic pronouncements about harnessing scientific developments for power projection during the first half of the century. These include predictions of physically enhanced troops carrying sensor arrays and implants giving them unmatched awareness of battlefields who will suffer reduced casualty rates when their injuries are repaired at the genetic level. While some of the proposed biotechnologies may never eventuate, others have already become commonplace, saving the lives of thousands of Western troops and offering the prospect that biotechnologically enhanced soldiers will become one of the new BW of the twenty-first century.

Engineers have already developed remote-controlled cyborg insects to conduct surveillance and possibly to deliver engineered genetic vectors. Research scientists, following in the footsteps of their 1920s counterparts, are promoting "more humane" weapons that will incapacitate their targets by altering their bodily functions through gene therapies delivered by particle-sized bullets. In this new environment, "the connotation and essence of military medicine" are poised to become directly associated with strategic offense, "and the goal of medicine is transforming from 'saving oneself and killing the enemy' to 'strengthening oneself and controlling the enemy'" (Guo, 2006: 1150, 1154).

As in the periods of the world wars and the Cold War in the last century, these biotechnological advances are being incorporated into the military force structures of the leading powers of the day with virtually no public debate of the kind that accompanied the advent of nuclear arsenals. Far more than the bacteriological weapons did, they hold a tremendous potential for ensuring the continued military and economic dominance of the states that wield them. Beyond the classical use of pathogens in warfare and terrorism, the development of various biotechnologies points toward a twenty-first century in which the most advanced states will attain offensive capabilities far beyond the capacity of less technologically advanced rivals or non-state actors.

This is not to say that the competitive advantages will last indefinitely or that the biotechnologies will not disseminate to other actors. While some of these developments were initiated by state military research programs, others have emerged from the private sector, which means that they might potentially be available to the highest bidders. In either case, many of the emerging technologies also promise to offer disparities of power as profound as those caused by the advent of atomic weapons in the last century.

THE BIOTECH REVOLUTION IN MILITARY AFFAIRS

The conventional wisdom on biotechnology has held precisely the opposite view, that coming decades will see Western nations increasingly vulnerable to ever-more-sophisticated WMD attacks, and that the acquisition of biological weapons by non-state actors and rogue states can, at best, only be slowed by constant intervention. As recombinant technology proliferates, a greater number of actors will possess genetic engineering capabilities that will enhance the lethality and durability of their biological weapons.

Competing Visions

Proponents of this perspective note that the overwhelming technological advantage in conventional forces enjoyed by the United States creates the incentive for competitors to develop effective asymmetric responses, and that the affordability, accessibility, and relatively easy preparation of biological weapons make them a likely means of doing so. In this view, the superior conventional capabilities of the US military not only fail to deter the proliferation of biological weapons but encourage their development. Western states will face an increasing number of biologically armed opponents and will remain on the defensive. The priority for military biotech research is therefore the development of protective equipment and vaccines, antibiotics, and antivirals (US Department of Defense, 1998, *Technical Annex*).

However, this scenario requires the presumption that military applications of biotechnology will simply be a secular progression from the bacteriological warfare that has existed throughout history. Even when analysts have factored in the vast possibilities of genetic engineering, it has usually only been to the extent that they can breed deadlier pathogens, and that the growing availability of technology means that it may be used by a broader spectrum of actors. Conventional wisdom therefore predicts an unstable future for the international system.

Yet rogue states have been mostly unwilling to use bioweapons, even when survival of the regime is threatened, as with Iraq, because they are subject to deterrent threats. Non-state actors such as the Aum Shinrikyo cult and the Amerithrax perpetrator have proven capable of launching biological attacks (at least domestically) and can generally not be threatened with deterrent military force because they reside within the state, and in some cases because they have suicidal intent. Yet their attacks are only as effective as the weaknesses in the defenses erected by nations that marshal far more resources than their assailants and are working constantly, if not always effectively, to reduce their vulnerability (see chapter 4). Rogue states, terrorist groups, and lone-actor psychotics are indeed relatively weak actors, which is both why the desire for an asymmetric advantage is imputed to them and why—barring the creation of an unstoppable synthetic pandemic—defensive efforts by well-equipped and alerted states will ensure that even genetically altered bacterial and viral agents will not pose serious threats to the survival of their regimes.

Instead, the leading states that are already the most capable actors in the international system will continuously integrate emergent biotechnologies into their military and national defense infrastructures and extend their dominance. This process will closely resemble the Revolution in Military Affairs (RMA) that occurred during the last thirty years of the twentieth century as the United States adapted its forces to exploit advances in new information technologies. The RMA, first described by Soviet military intelligence in the 1970s and then witnessed by the world during the unexpectedly uneven 1991 Gulf War, occurred because the United States employed its competitive advantage in integrated computer systems. Rather than a single transformative device, like the atomic bomb, the steady accretion of advanced technologies augmenting existing equipment came to inform doctrine and strategies.

The term *asymmetric warfare* is meant to describe efforts by weaker participants in military confrontations to frustrate the advantages of the stronger power by guerilla tactics or other unconventional methods not envisioned in force planning (Mack, 1975). However, high technology also offers asymmetric advantages to the best-equipped actors, and American military planners sought to use the advances of the RMA to field forces that no state competitor could match. Their goals included "dominant maneuver" capability on the battlefield in bringing dispersed resources to bear against targets, "precision engagement" capability delivered by smart weapons, and "full dimension force protection" against all anticipated threats (Rizwan, 2000). The ultimate expression of this vision would be a fighter comparable to a Jedi knight from the *Star Wars* films: a super-empowered soldier, dressed in a protective stealth cloak and commanding an armed companion unmanned aerial vehicle (UAV), able to perform solo missions and transmit data back to headquarters (RAND, 1994).[1]

RMA adaptations are also driving the imperative of "maintenance of the human weapon system." Arguing that critical precision missions are now performed by single pilots rather than the hundreds required for bombing runs during World War II, one US Air Force base commander wrote of "our duty to optimize the performance of our Airman through fitness and wellness" (Alder, 2009). With such advancements in hardware, planners could be forgiven for describing humans as "the weakest link" as Director Goldblatt did on the eve of a new wave of Pentagon biotechnology R&D.

However, military planners likewise foresee similar advantages conferred by developments across the various biotech fields beyond human performance enhancement. In coming decades, biotechnology is forecast to bring advances such as "rugged computers" made from biological components that will provide situational awareness to individual soldiers on the battlefield, camouflaged materials and lightweight armor incorporating the properties of living organisms, and ingested biological markers to distinguish friendlies, which would be of particular use in counterinsurgency (Purdue University, 2001a). From the perspective of those reimagining force planning, the anticipated future is not one of vulnerability but unassailability.

Defense R&D

While some military (or potentially military) applications of biotechnology are products of the private sector (discussed later in this chapter), it is no state secret that militaries actively sponsor their own biotech R&D programs. The US Department of Defense in particular is open about the large number of such projects that it oversees. And while most of these are described as intended for troop protection, many are clearly intended to enhance combat operations. As with pathogen stockpiles maintained ostensibly for defensive research, it is only the intent of the wielder that determines whether or not they are offensive. And, as the Amerithrax case illustrates, even projects officially intended for defensive purposes may be misapplied.

Much of the research is conducted under the auspices of the Pentagon's Defense Advanced Research Projects Agency (DARPA) rather than through legacy programs remaining from the era of bacteriological weapons stockpiles. Military transformation analyst P. W. Singer has described DARPA biotech projects as attempts to make soldiers "kill-proof" (Lin, Mehlman, and Abney, 2013: 9).

Established in 1958 as a response to the launch of the first Sputnik satellite by the Soviet Union the year before, DARPA was intended to promote "high-risk, high-payoff" R&D in areas beyond the immediate envisioned needs of military planners. The agency's singularly most influential project has undoubtedly been a communications system that came to be known as ARPANET before penetrating—and transforming—the commercial sector as the Internet.

One of the prime movers on this project, J. C. R. Licklider, director of DARPA's Information Processing Techniques Office, also laid the groundwork for later biotech developments. In addition to calling for the development of networked personal computers and virtual-reality programs, in his 1960 call for "Man–Computer Symbiosis," Licklider "foresaw using new types of computational capabilities to achieve, first, augmented human capabilities, and then possibly artificial intelligence" (Van Atta, 2008: 20, 23, 27). DARPA began work in 1973 on brain–computer interfaces that are used for contemporary neurotech research into soldier human performance improvement (Miranda et al., 2015: 54). DARPA support for biotech projects increased sharply under Anthony Tether, agency director from 2001 to 2009, but in response to criticisms by the President's Council on Bioethics and some members of Congress that DARPA was creating a "Frankenstein army . . . program names were changed to dull their mad scientist edge" (Shachtman, 2007).

While a reported 90 percent of its projects fail to come to fruition, high-profile DARPA research that has had a significant impact on US military capability includes Saturn rockets, ground radar, stealth fighters, Predator missiles, and drones. The agency's budget of $3 billion is small compared to intelligence agencies, but it supports an "open culture" promoting "radical innovation" praised by participant scientists, most of whom are university researchers (Basken, 2013; Moreno, 2006: 12–13).

Indeed, "DARPA itself does not conduct scientific research" but sponsors it in research laboratories (Miranda et al., 2015: 53). One DARPA research partner is the Institute for Collaborative Biotechnologies (ICB), "an Army-sponsored University Affiliated Research Center . . . led by the University of California, Santa Barbara, in collaboration with the Massachusetts Institute of Technology, the California Institute of Technology and partners from the Army and industry" (Regents of the University of California, 2013). Even when particular DARPA programs end, their lead researchers often carry on their work for other organizations, including private corporations, but building on advancements funded by taxpayers through the Pentagon (Dewar, 2015). Future secretary of defense Ashton Carter (2001: 17, 157–158), in anticipating the preeminence of biotechnology in future military operations and noting that the military would be unlikely to retain top scientists in-house, called for deepening ties with the commercial research sector, other government agencies with longer-standing relations with the biotech sector, and the establishment of "a university-affiliated, government-owned laboratory" system for conducting relevant R&D.

This arrangement is reminiscent of the networked R&D structures employed by the twentieth-century military biotech programs in other countries described in chapter 1. The traditional BW programs of both superpowers drew upon expert exchanges and partnerships with researchers from academic institutions and relevant industries (Ben Ouagrham-Gormley, 2014: 40, 70).

In 2014, DARPA announced the creation of a new Biological Technologies Division, built from existing research units and new programs, intended to ensure that biotechnology is not merely an aspect of various research programs but that "biology takes its place among the core sciences that represent the future of defense technology." The new division's objectives are, first, to "restore and maintain warfighter abilities," to "harness biological systems" in part by creating new biological material in its Living Foundries program, and to "apply biological complexity at scale" by understanding the interactions of mammals with other organisms "to enhance global-scale stability" (DARPA, April 1, 2014).

Similarly, the Pentagon Office of Net Assessment (ONA), which envisions potential future strategic environments and challenges, has also promoted biotech R&D as a defense priority. In 2002, the ONA recommended revising federal regulations to allow experimental biotechnologies to be brought to the battlefield more readily. It also called for facilitating a greater partnership with private-sector researchers by restricting antitrust laws to permit quicker product development (Armstrong and Warner, 2003).

In 2004, the Department of Defense established the Human Performance Resource Center with the mission to "improve the human weapon system's ability to accomplish the mission." Its research priorities included "sleep management, performance at altitude and in the heat, workout regimes, energy management, social/psychological resilience, and decision-making/cognitive enhancements" (Kotch, 2010).

The Army Research Laboratory also organizes biotech R&D related to human performance enhancement for soldiers. In a 2012 presentation, its Soldier Performance Division chief stated his organization's multiple goals, not all of which related to tending to injured warfighters: "Testing on soldier perception and cognition is expected to produce the payoff of increased situational awareness, lethality and survivability." He went on to describe cognitive neuroscience as a "top three funding priority" for the White House, and one of the "top six potential disruptive basic research areas" for the US Army, noting that the neurotechnology sector was investing over $140 million annually in defense research and had produced more than a half million citable documents on "Army-relevant neurotechnology applications" since 1998. However, to achieve the Army's future augmentation goals, "solution requires we move beyond the laboratory setting" (Lockett, 2012: 11–13).

THE FUTURE IS NOW: FROM LAB TO BATTLEFIELD

Whether in collaboration with the private sector or directly from their own research facilities, leading state military programs are implementing biotech innovations that have already had significant strategic impacts beyond the realistic aspirations of non-state actors. The lives of thousands of Coalition troops have been saved by biotechnologies deployed in Iraq and Afghanistan, and other projects being implemented will enable soldiers to fight more effectively under more adverse conditions than previously possible. In short, rather than being curtailed by asymmetric defenses, the power projection capabilities of the strongest actors in the international system will increase markedly during the twenty-first century.

Troop Health and Survivability

Despite the mechanization and increased destructive power of warfare in the mid-nineteenth century, it was not until World War II that wartime combat deaths exceeded those that occurred subsequently off the battlefield. This shift, due to advances in combat medicine, permitted American and British forces to conduct forward operations with reduced fatalities. Advances in biotechnology are responsible for the continuation of this trend into more recent conflicts with similar results (Frank, 2007).

When the United States invaded Iraq in 2003, many of its soldiers and marines were treated with $90 bandages produced by an Oregon-based company called HemCon Inc. The military ordered twenty-six thousand of these dressings, made from a shrimp shell extract called chitosan, which stopped arterial bleeding within a minute of application to wounds. Another bandage, developed by the American Red Cross but with limited use because of its $1,000 price tag, was made of clotting proteins extracted from human blood. By contrast, a powdered coagulant manufactured by Z-Medica called Quik-Clot that could be poured directly onto wounds was issued in first-aid kits, initially to marines and then across service branches. QuikClot is a granular substance that can be poured directly onto a wound, almost instantly forming a clot that stops bleeding. A hemostatic agent in QuikClot draws water molecules out of blood and promotes accelerated clotting (Allen, 2003). It also generated both a heat-giving exothermic reaction and complaints by recipients that they were receiving burns as a result, sometimes making it difficult for doctors to remove damaged tissue. By 2010, both the US Army and Marines had switched to providing QuikClot combat gauze in first-aid kits instead (Cavallaro, 2010). Other options are available as well:

> Biological materials are now known that have excellent adhesive properties and can help stop bleeding. These include adhesives from barnacles. . . . Biosealants with excellent adhesive properties might be developed (e.g., by modifying protein biopolymers), and individual soldiers might carry them in their backpacks. The biosealant would act as a "super glue" to stop bleeding and hemorrhaging until the injured soldier could be evacuated to a more permanent treatment setting. (Committee, 2001: 36)

More controversially, military doctors in Iraq injected more than one thousand wounded troops with the experimental coagulant Recombinant Activated Factor VII. The federal Food and Drug Administration (FDA) had only approved it for treating rare forms of hemophilia and had cautioned that it could produce fatal blood clots in the lungs, hearts, and brains of patients with normal blood (Little, 2006). The drug, marketed as NovoSeven, had first been used to treat a wounded Israeli soldier in 2002 but was quickly adopted by other militaries. The British Ministry of Defence defended the practice in 2006: "It has only twice been administered—in the two separate incidents in Iraq, and on both occasions the individuals' lives were almost certainly saved by the treatment. It is used only in extremis, when the casualty has suffered a catastrophic trauma, and when no other treatments are viable or available" (Dyer, 2006). Subsequently, DARPA has tasked partner company Arsenal Medical with developing its hemostatic foam into a product that could be used to stop internal bleeding even without direct access to the combat wound (DARPA, "Wound Stasis System," 2013).

The use of biotechnologically advanced coagulants to treat severe combat injuries had a substantial effect on the first wars of the twenty-first century: "The ratio of [American] combat-zone deaths to those wounded has dropped from 24 percent in Vietnam to 13 percent in Iraq and Afghanistan. In other words, the numbers of those killed as a percentage of overall casualties is lower." However, while fatalities are down, the continuation of casualties has meant bearing the increased costs of more survivors with amputations and psychological damage (Knickerbocker, 2006): "In recent years, veterans receiving mental health care from the Department of Veterans Affairs (VA) have constituted almost a third of the total number of veterans receiving health care from the VA" (Miranda et al., 2015: 64).

Still, because reductions in public support for wars are attributed to high fatality rates, contemporary military planners are more interested than ever in minimizing costly operations. By the time of the War on Terror, 55 percent of battlefield deaths were due to excessive blood loss (Armstrong and Warner, 2003). But the development of rapid coagulants sharply reduced the rate of combat deaths and may have sustained the American public's tolerance for the Iraq and Afghanistan missions. The potential implication is that democracies may become more willing to engage in future wars if the human costs of doing so are minimized.

But while technological advancements in battlefield medicine and armor "gave troops a better chance of coming home than any other generation of war fighters," this does not mean that they have escaped unscathed. Nearly half of Afghanistan and Iraq veterans from the 2000s have filed disability claims, more than double the number who did so after the 1991 Gulf War (Chandrasekaran, 2014). The increased social costs of wounded and disabled veterans would require other investments.

Biotech is being employed along these lines across a variety of projects: "Technologies are under investigation to fully restore complex tissues (muscle, nerves, skin, etc.) after traumatic injury and, most dramatically, to develop neural-controlled upper extremity prostheses that fully recapitulate the motor and sensory functions of a natural limb" (DARPA, "Restorative Biomedical Technologies," 2010). Research of this type is progressing along an array of fronts.

In the area of combat medicine, DARPA is moving beyond coagulants. Its projects involve blood "pharming" that will produce engineered red blood cells (DARPA, "Blood Pharm," 2010) that can be preserved for delivery to the front lines to enable transfusions for wounded troops despite the "austere conditions of forward deployment" (DARPA, "Long-Term Storage of Blood Products," 2010). A related program would use hormone therapy to extend the survivability of combatants losing critical amounts of blood before fluids and transfusions can reach them: "Achieving this goal will allow increased time—perhaps many hours or even days—for evacuation, triage, and initiation of supportive therapies" (DARPA, "Surviving Blood Loss," 2010).

To better treat other battlefield wounds and reduce rehabilitation needs, "DARPA seeks to create a dynamic putty-like material which, when packed in/around a compound bone fracture, provides full load-bearing capabilities within days, creates an osteoconductive bone-like internal structure, and degrades over time to harmless resorbable by-products as normal bone regenerates" (DARPA, "Fracture Putty," 2010). In vivo biomaterials, or compounds directly incorporated by a living organism, would regenerate tissue and then be fully absorbed (Armstrong and Warner, 2003). An additional development that could reduce infection and mortality among burn victims is a "self-medicating" bandage. "Laced with nanoparticles, it detects harmful bacteria in a wound and responds by secreting antibiotics" (De Lange, 2010).

A fully functional prosthetics program termed HAND "is developing the fundamental research that will enable the use of neural activity to . . . restore natural function through assistive devices. By directly harnessing the ability of neural pathways to operate natural systems, the HAND program seeks to provide means of restoring the lives of injured warfighters" (DARPA, "Human-Assisted Neural Devices," 2010; DARPA, "Reliable Neural Interface Technology," 2013). In 2013, DARPA-funded research enabled amputees to experience partial sensations of touch through prosthetic limbs via neural interfaces (DARPA, "New Nerve and Muscle Interfaces and Wounded Warrior Amputees," 2013). In 2015, DARPA showcased a prosthetic arm that not only gave the user realistic touch sensations but assisted him in climbing a rock wall prodigiously (Zhang, 2015). That such research is being conducted by the Pentagon speaks to its perceived strategic value.

In 2015, the US Department of Veterans Affairs agreed to pay the full $77,000 costs of exoskeletons that enable locomotion for paraplegic veterans. The ReWalk system consists of "leg braces with motion sensors and motorized joints that respond to subtle changes in upper-body movement and shifts in balance" (Watson, 2015).

This line of research connects with the 2003 DARPA Strategic Plan, coinciding with Director Goldblatt's speech, which called for "turning thoughts into acts" to create "U.S. warfighters that only need use of the power of their thoughts to do things at great distances." Enabling the human brain to directly control a peripheral device such as an artificial limb also means that it could control robots on the battlefield that could fight without risk to soldiers (Moreno, 2006: 9, 39). Such avatars would be the infantry equivalent of drone aircraft and would profoundly change the nature of soldiering more than UAVs are now doing with piloting.

Other developments in preventing or restoring injuries to troops are more overtly related to battlefield performance. DARPA, in noting that "the negative impact that stress has on the cognitive, emotional, and physical well-being of warfighters is irrefutable," proposes that "novel molecular biological techniques, coupled with in-vivo measurement technologies, can allow for management of the stress pathways and behavioral analysis in real time" (DARPA, "Enabling Stress Resistance," 2010). Biodegradable, self-regulating drug-delivery systems will deliver precise therapeutic or performance-enhancing pharmaceuticals in combat (DARPA, "Feedback Regulated Automatic Molecular Release," 2010). Other projects seek to prevent brain damage caused by the kinetic force of explosives (DARPA, "Preventing Violent Explosive Neurologic Trauma," 2010) and to develop REMIND, "a neural prosthesis for lost cognitive function and memory impairment" for troops with traumatic brain injury (DARPA, "Restorative Encoding Memory Integration Neural Device," 2010).

Advances in 3-D printing have already permitted living stem cells to be "loaded into a printer and spat out in new shapes, including in three-dimensional tissues and structures . . . for the regeneration of muscle, cartilage, and even nerves . . . once thought to be an unreachable goal." In 2014, a Polish firefighter recovered sensory and motor function after his fully severed spinal cord was patched with a 3-D-printed stem cell bridge, and 3-D-printed rib cages for lung cancer patients and other such prosthetics are already being deployed in hospitals (Pash, 2015).

Another option is using engineered viruses for delivery of genes. In 2005, researchers were able to restore enough damaged cochlear hair cells in deafened guinea pigs to recover 50 to 80 percent of their hearing levels. The therapy was delivered by adenoviruses engineered both to render them harmless and to produce a hair growth stimulant (Coghlan, 2005).

> Gene therapy agents could be transfected into cells by bombarding a patch of skin with DNA-coated pellets from a gene gun. As the cells are sloughed off, expression of the therapeutic protein would naturally cease but could be renewed by another application of the agent. By 2025, reliable and robust means of delivering DNA constructions to other cell types will also become available. In fact, much or all of the technology implanted into the individual soldier will probably be derived from the individual's own cells rather than from fabricated devices. (Committee, 2001: 70)

Researchers affiliated with ICB are developing synthetic viruses that can "hijack the replication machinery of the cell to invade the host species" to deliver drugs against cancer cells (Soh, 2013). While the benefits to humanity of such research are obvious, the direct benefits to the US Army that funds the research are not—without extrapolation.

Another possibility is neural or cortical implants such as prosthetic retinas, both to treat injuries and to offer enhanced abilities. "As the risks and costs associated with neural implants are reduced, they may be used to increase the visual and hearing acuity of unimpaired individuals to levels well above average. Soldiers possessing these extraordinary faculties would be well suited to gathering intelligence and performing long range reconnaissance missions" (Committee, 2001: 38). Shortly after the announcement of the Microsoft HoloLens virtual-reality gaming technology, DARPA mooted adapting the technology to serve as a replacement for eyes lost to battlefield injuries, or to provide enhanced vision in combat (Joslin, 2015).

Another program with the goal of "enhancing combat performance" studies the influences of biological clocks on soldier health (DARPA, "Biochronicity," 2013). In 2014, the Obama administration announced its priorities for improving Veterans Administration care. Among them was calling for all service members to eventually be fitted with miniaturized versions of DARPA's prototype electrical prescriptions (ElectRX), "new computer chips designed to modulate the nervous system to help with everything from arthritis to post-traumatic stress" (Lamothe, 2014).

Human Enhancement

Beyond treating wounded soldiers, DARPA researchers note that there are also tremendous "commercial BCI applications for healthy individuals" (Miranda et al., 2015: 53).

It also seems likely that the first impact of advanced biotechnologies on the battlefield will manifest in this area. As Lele (2009: 141) notes, at present most military biotech R&D is focused on "improving the material of war, enhancing the performance of warriors, and using biological processes to improve systems design and performance" rather than in offensive systems that directly degrade enemy capabilities.

Efforts to field augmented troops represent new approaches to the use of biotechnology in warfare, a qualitative shift away from traditional but uncertain bacteriological weapons to entirely new strategies for assuring battlefield dominance. As proponents of this biotech approach envision it, "futuristic, 'superhuman' capabilities of individual soldiers could enable small units to operate for extended periods of time, carry the fight to remote locales, and endure harsh extremes of climate" (Committee, 2001: 7). Moreno (2006: 114) argues that "the first state (or nonstate) actor to build a better soldier will have taken an enormous leap in the arms race." Solider augmentation also offers the potential for obviating difficulties in military recruitment that the United States has experienced post–Iraq War that has compelled it to ease standards for enlistment, including lower standards of physical fitness and higher age ceilings (Shachtman, 2007).

Although seemingly fantastic, billions of dollars have already been spent on several programs directed toward fielding various types of Augments.[2] As with biotechnologies to increase survivability, fielding mechanically or biologically augmented soldiers offers multiple benefits for states with the capacity to do so. It also raises a host of political and ethical questions without clear answers. Certainly, there would be tactical advantages for militaries whose personnel are able to operate more effectively than their adversaries under difficult conditions. And the boon of losing fewer servicemen to injury, and being able to return those who are injured to the front lines more quickly, is evident. But there are also broader potential national and international political impacts. Democratic governments, which endeavor to avoid costly or risky wars (Gartner and Segura, 1998; Reiter and Stamm, 2002), might be tempted to exert their power as their conventional force advantages grow, and as the costs of providing for disabled veterans diminish.

Great powers with both conventional and asymmetric biotechnological edges over rivals may be open to the use of force to maintain their positions if they are secure in the knowledge that they are well beyond the capabilities of opponents to match them. The advent of nuclear weapons is credited with reducing the number of interstate wars, with the effect of entrenching the hegemony of the technologically advanced states that wield them. RMA advances gave the United States such a lopsided advantage in its early post–Cold War interventions (zero combat deaths during nearly three months of NATO missions during the Kosovo War), and its initial easy success in toppling Saddam Hussein from power in Iraq led, temporarily, to rapprochement efforts by "rogue" regimes in Iran and Libya to avoid the same fate. While advanced equipment is responsible for these successes, biotech now offers the opportunity to enhance the performance of the individual combatants themselves.

[DARPA] is engaged in the development of designer drugs that will increase cognitive functioning, including attention span and alertness after periods of sleep deprivation. Another area for future research is "neural prostheses" that will enable commanders to monitor the vital signs of soldiers in the field or even to permit the control of UAVs directly by pilots in remote locations. (Huang and Kosal, 2008)

In 2002, DARPA launched the Augmented Cognition (or AugCog) initiative, a project dedicated to developing a headband that monitors brain activity. Among the objects is to determine if military personnel in the field are receiving too much sensory input to process effectively, and send alternative information instead. A 2005 trial of the device resulted in subjects doubling their recall, and improving 500 percent in working memory. (*The Economist*, February 25, 2010)

Research on reducing the amount of sleep that soldiers and pilots require to function effectively has become a global enterprise, with countries including France, Canada, Singapore, and Taiwan establishing military research units in this area. In the language of these projects, fatigue and even sleep are described as operational weaknesses preventing humans from taking full advantage of their equipment, weaknesses that intervention can ameliorate. Some major powers have already begun the attempt: during the Iraq War, the British Ministry of Defence purchased twenty-four thousand tablets of one of the most promising drugs, modafinil, and the United States and France both began to routinely supply it to pilots. The use of stimulants by militaries is so widely entrenched, with amphetamines in regular prescribed use for decades (Saletan, 2013, and see World War II section of chapter 1), that Savulescu and Bostrom (2009: 2) question whether the use of modafinil is qualitatively different from "a good cup of tea." But the premise of reducing or eliminating the need for sleep as a component of troop health is a recent development.

Additionally, DARPA has provided congressional testimony about its Continuous Assistance Program (CAP) that would "make the individual warfighter stronger, more alert, more endurant, and better able to heal . . . prevent fatigue and enable soldiers to stay awake, alert, and effective for up to seven days straight without suffering any deleterious mental or physical effects and without using any of the current generation of stimulants." Potential approaches include the use of transcranial magnetic and electrical stimulation to activate brain pathways and to enhance learning (Moreno, 2006: 11, 118). Research to enhance soldier training began at the University of New Mexico with DARPA sponsorship in the 2000s (Temple-Ralston, 2015).

In the meantime, the military relies on more conventional stimulants, the pitfalls of which have received significant attention. B-1 bomber pilots who operate nineteen-hour flights between the Persian Gulf and the United States take Dexedrine, an amphetamine known as speed or "go pills." One such pilot, who subsequently went drinking with buddies before attacking them in a fit of paranoid delusions in which he seemed to believe he was in the television series *24*, was acquitted by a court-martial after military psychiatrists concluded he suffered from a "substance-induced delirium" (Murphy, 2012). American pilots who killed Canadian soldiers in a 2003 friendly-fire incident in Afghanistan had also been on Dexedrine during thirty-hour missions (Moreno, 2006: 115).[3]

Another DARPA neural program with battlefield applications is Silent Talk, which would develop the capability to communicate without speaking by recognizing the neural signals for specific words. Linked devices would permit troops in the field to recognize the signals for the "intended speech" of at least one hundred words commonly used by troops in combat operations (Warwick, 2009). Beyond the advantages of silent communication and preventing hostile forces from intercepting messages, such technology would effectively produce electronic telepathy and have a tremendous commercial-sector potential for hands-free communication.

While Augments would be able to receive greater amounts of situational information on the battlefield through neural devices, processing it effectively is another matter. Technologies developed through the AugCog and Enabling Stress Resistance projects might alert commanders that individuals are suffering mental or physical exhaustion. Another approach would be to "develop quantitative and integrative neuroscience-based approaches for measuring, tracking, and accelerating skill acquisition and learning while producing a twofold increase in progression in an individual's progress through stages of task learning." Reminiscent of the neural training uploads for particular weapons systems and martial arts in the science-fiction *Matrix* films, results would be achieved through the "development of neurally-based techniques for maintenance of acquired skills [and on] preferential brain network activation" (DARPA, "Accelerated Learning," 2010).[4] In testing the Accelerated Learning program, "rifle marksmanship training was used as the militarily relevant task" (Miranda et al., 2015: 61). "Within the ICB Cognitive Neuroscience task order, decision-making research also considers the role of sensory noise and target detection" (Grafton and Miller, 2013).

Other biotechnologies would provide physical enhancements to Augments. The field of biomimetics seeks to mimic useful naturally occurring characteristics in living organisms. For example, ants and spiders can lift loads dozens of times their own weight, and horses can withstand freezing temperatures without thick hair. "Understanding how horses and other animals overcome drastic changes in their environment would be extremely useful. As a measure of the importance of biomimesis, the Army has declared biomimetics one of its Strategic Research Objectives (primary focus areas for basic research)" (Committee, 2001: 14–15).[5]

In her 2012 TED talk, Director Dugan described a DARPA project to develop a hypersonic aircraft to fly at MACH 20 to reach any point on the globe in under one and a half hours. Such an aircraft would require extraordinary maneuverability, so DARPA began to study the aerodynamics of the most maneuverable bird in nature, the hummingbird. DARPA then began work on a drone that looked and flew just like a real hummingbird, including backward and in rotation, and weighed less than an AA battery despite being equipped with a video camera.

Another project already at least at the prototype stage utilizes an electrically charged undersuit "focusing on the soft tissues that connect and interface with the skeletal system." The goal of Warrior Web is "augmenting the work of Soldiers' own muscles to significantly boost endurance, carrying capacity, and warfighter effectiveness" (DARPA, "Warrior Web," 2013; DARPA, "Warrior Web Prototype Takes Its First Steps," 2013).

Power Projection

Unless the R&D invested in these projects proves futile, the US Department of Defense is indeed on its way to developing not just super soldiers, but essentially comic book superheroes. One $3 billion program, begun in 2002, is intended to create a "metabolically dominant soldier" who will be enabled by gene therapy to lift up to eight hundred pounds, block pain receptors for days, and "run at Olympic sprint speeds for 15 minutes on one breath of air" (Sokolove, 2007).

And if neural or cybernetic prostheses and gene therapy do not produce a Superman—or Captain America—the contributions of other research programs may still permit the fielding of a biomimetic Spiderman. Or, as Director Dugan (2012) put it, "maybe Spiderman will one day be Geckoman":

> The Z-Man program will develop biologically inspired climbing aids that will enable an individual soldier to scale vertical walls constructed of typical building materials without the need for ropes or ladders. The inspiration for these climbing aids is the way geckos, spiders, and small animals scale vertical surfaces. . . . The overall goal of the program is to enable an individual soldier using Z-Man technologies to scale a vertical surface while carrying a full combat load. (DARPA, "Z-Man," 2010)

In 2014, DARPA revealed a prototype that allowed for "a 218-pound climber ascending and descending 25 feet of glass, while also carrying an additional 50-pound load in one trial, with no climbing equipment other than a pair of hand-held, gecko-inspired paddles" (DARPA, June 5, 2014). Concurrently, ICB researchers undertook a project titled Design Principles and Strategies for Biomimetic, Gecko-Like Ambulation (Israelachvili, Turner, and Byl, 2013). Nowhere on the project web page does it mention military applications of "climbing robotics," but the category page above it states that "the aim of the ICB Control and Dynamical Systems task order is to adapt the myriad of highly optimized sensing, actuation, and control strategies discovered by nature to the future needs of the soldier" (Hespanha and Plaxco, 2013).

Endowing troops with the ability to scale vertiginous surfaces obviates classical applications of biowarfare: no need to flush out the enemy by hurling plague-ridden corpses over fortifications when you can simply walk up them instead. This is perhaps the most outlandish example of how biotechnologies are being developed to aid military power projection capabilities, but it is by no means the only one. Another completed biomimetic project increased the efficiency of human swimmers by 80 percent and more than doubled their speed by giving them oscillating foils based on the propulsion mechanisms used by some fish and sea birds (DARPA, "PowerSwim," 2010).

A project to achieve rapid altitude and hypoxia acclimatization would permit the fielding of troops (perhaps in potential battle zones such as the Hindu Kush or the Himalayas) with "novel pharmacological, biological, and technological approaches to adapt to high altitudes (4,000–6,000 meters)" (DARPA, "RAHA," 2010). This research was likely the result of DARPA-supported studies that demonstrated that hydrogen sulfide gas could increase survival time from minutes to hours in mice suffering from both hypoxia and blood loss. Another study found that gloves that raised and lowered the temperature of blood circulating within the body could permit humans to survive hypothermia and to remain vital in extremely hot conditions with the right field gear. The cooling gloves were also performance enhancers against muscle fatigue, permitting the project director to increase his exercise regimen from one hundred to six hundred pull-ups in six weeks and to perform one thousand push-ups on his sixtieth birthday (Shachtman, 2007).

The adoption of biotechnology to enable force projection began during the colonial era, when Europeans discovered that quinine could prevent malaria, thus opening the door for the Scramble for Africa. Shortages of antimalarial drugs during World War II caused such high morbidity rates among American personnel serving in the Pacific that General Douglas MacArthur remarked that the campaign would be a slow one unless additional measures were taken (Marble, 2010).

In the twenty-first century, antimalarial drugs remain a challenge to force projection. Mefloquine, a comparatively affordable antimalarial also marketed as Lariam, has severe psychiatric side effects, first noted in the Vietnam War. Problems include psychotic behavior, paranoia, and hallucinations. A 2003 report indicated that "Mefloquine use was a factor in half of the suicides among troops in Iraq in 2003—and how suicides dropped by 50 percent after the Army stopped handing out the drug." Its use was also linked to murders and suicides by special forces personnel in Afghanistan between 2002 and 2004 (Benjamin, 2012). In 2012, Roche Pharmaceuticals, the maker of Lariam, notified the federal Food and Drug Administration that it had been alerted by a physician that a patient with traumatic brain injury taking the drug, presumed to be a serviceman, had been charged in a high-profile massacre of civilians involving seventeen victims (Ritchie, 2013).

When difficulties with malaria mounted during the Vietnam War, including transmission back to America, the US Navy utilized recombinant technology to develop a DNA vaccine to prevent malaria infections. When the program began in the 1990s, the majority of troop deployments were to malarial regions, and the *Plasmodium* parasites were the top cause of casualties in Somalia. In tests announced in 1998, research teams were immunized with *Plasmodium* DNA, with the majority of participants developing T cells that function as antibodies when confronted with malarial parasites. This development involved the creation of malaria vectors that functioned like common vaccines, potentially opening the way to safer deployments for American marines (Gillert, 1998).

The advent of DNA vaccines of this type theoretically allows scientists to develop vaccinations against all known diseases. The Naval Medical Research Institute therefore created a "phage library" for the purposes of developing antibodies to all possible strains of infectious agents (Wang, 1998). As the technology is further developed, the militaries of advanced states will increasingly turn to active biotech solutions to biological threats, rather than pharmaceutical prophylactics. However, with defense planners concerned by the possibility of the use of genetically modified bioweapons by rogue and non-state actors, they will also conduct further research into countering genetically engineered vectors that might be created to replace the naturally occurring agents against which American forces are already protected (US Department of Defense, 1998, *Technical Annex*). DARPA's "7 Day Biodefense" (2013) program seeks to develop persistent and transient immunities to unspecified pathogens out of the recognition that unfamiliar agents would likely be employed in a major biowarfare attack. Its subsequent "Folded Non-Natural Polymers with Biological Function (FOLD F(x))" (2015) intends to analyze and catalog all polymers in the human body, for the stated purpose of being able to have ready defenses against novel biological threats.

Accompanying this effort, and parallel civilian homeland security programs described in the next chapter, are programs to "improve the speed, efficiency, yield, flexibility, and expense of current-generation vaccines and monoclonal antibodies." Military planners determined that bulk production of vaccines using eggs, the same practice employed in the creation of Cold War pathogenic weapons, is vulnerable to contamination and hampered by the "inability to grow certain dangerous viruses in eggs." The new program would be adaptive; instead of stockpiling countermeasures for currently recognized threats, it would aim to provide "on demand" production of three million doses of vaccine-quality recombinant protein and monoclonal therapies in a three-month time frame (DARPA, "Accelerated Manufacture of Pharmaceuticals, AMP," 2010). Obviously this strategy carries the risk that such a period of time would be insufficient to develop treatments and distribute them to exposed troops on forward deployments and large numbers of civilians at home.

Genetically Modified Foods

The apparent end of the use of industrial equipment for military biotech purposes does not portend the resolution of the dual-use dilemma. As the planners of the AMP project note, it is still necessary to work with deadly pathogens if one is to find treatments for them. Another commercial-sector field that is experiencing securitization, and is already highly controversial in its own right, is genetically modified food. Called "Frankenfoods" by their vocal detractors and genetically modified organisms (GMOs) by agribusiness, they represent a growing number of plant and animal products that have been the recipients of recombinant engineering to, among other results, increase their yield, improve their flavor, or lengthen their shelf lives.

GMOs potentially hold a number of possibilities for military purposes. As far back as 1960, the US Air Force and Navy funded studies to determine whether ions accelerated plant growth and could thereby feed troops on forward deployments (Krueger et al., 1962). More recently, the US Army has initiated programs to develop crops with enhanced levels of nutritional components, built-in vaccines, or edible factors that impart resistance to spoilage (Committee, 2001: 53).

In particular, "functional foods" are expected to reduce logistical demands, which would enable more efficient power projection. Such foods have been modified to provide more than their normal nutritional value. Instead, they can contain nutraceuticals, "such as naturally occurring anti-microbials that inhibit certain pathogens known to exist in a given operational area. Or foods could be designed with vaccines in them, and an army could be vaccinated quickly and efficiently by distributing genetically engineered food" (Armstrong and Warner, 2003).

In the seemingly mundane and unappealing area of field rations, significant developments in food processing will enhance flavor and nutrients and dramatically extend shelf life. In particular, stabilized water activity control can preserve fresh bread in MREs for up to three years. Emergent fermentation technology will provide large quantities of engineered high-nutrient proteins grown in chemical tanks that can be made to taste like nearly anything. Such extended life span and portable production capability for food will greatly reduce current logistical demands, augmenting force mobility and reducing dependence on regular supplies (US Army, *Sustained Food Quality*, 1998). Genetically modified food is also being developed to be highly digestible to reduce the quantity of rations that require transportation and with biomarkers to distinguish the ingestor as a friendly under battlefield or peacekeeping scans (Egudo, 2004: 14).

Biofuels

Researchers are also seeking to engineer plants and animals to provide another essential resource for military power projection, namely energy. If a truism behind the development of GMOs for enhanced food is that armies fight on their stomachs, it is also recognized that it is conflicts over access to fuel supplies that send them into battle. From Japan's decision to launch a preemptive strike against the United States in 1941 after the enactment of oil sanctions that would cripple its war machine to Iraq's invasion of Kuwait in 1990 over accusations of slant drilling across their border, the modern era has seen the shedding of a great deal of blood to preserve access to petroleum. A lack of access to fuel has also hobbled the efforts of forces in the field. The German Sixth Army ultimately capitulated at Stalingrad when its fuel supplies were exhausted, and American forces under General George Patton advanced so quickly through France that they exceeded the ability of their supply lines to refuel their vehicles (Purdue University, 2001b).

One solution is to reduce dependence on fossil fuels and generate energy directly from living or recently living organisms, or biomass. Most proposals for using plants and wood as alternative fuel sources require the capability to extract ethanol from them. DARPA oversees several projects to harness and increase the efficiency of fuel from biomass for strategic purposes, including as a food source for troops on forward deployments (DARPA, "Crystalline Cellulose Conversion to Glucose, C3G," 2010).

The ability to convert from a petroleum-based to biomass-derived ethanol fuel economy carries tremendous consequences for the stability of the international system. It would mean a sharp reduction in the importance, and presumably the cost, of oil for the first time since the nineteenth century. The results would potentially be tremendously destabilizing for existing major energy producers such as Russia and OPEC states.[6] Rogue or eccentric petro-state regimes, such as those of Iran and Venezuela, have already been buffeted by social unrest in response to government service cutbacks and unemployment related to the decline of petroleum prices stemming from the 2008 global economic recession. It is impossible to predict whether the pressure from further diminutions of the fossil fuel industry would lead to popular democratic regime transformation in these states or would prompt embattled rulers to tighten their grip and threaten regional wars to consolidate popular support, as indeed both Tehran and Caracas did.

For fossil fuel net importers, a category that the United States joined in the decades after Pearl Harbor, the shift to biomass means the potential for self-sufficiency.[7] Brazil, for example, has rapidly become a twenty-first-century global leader, and one of its signature accomplishments has been the attainment of energy independence. While it continues to use petroleum, by 2007 roughly half of the energy supplied to the population of nearly two hundred million came from sugarcane ethanol, including fuel for cars (Reel, 2006). Dependence on imported fuel significantly constrains policy options and power projection capabilities, as Japan realized in 1941. But energy independence and the capacity to become a net exporter can confer or restore domestic power and international influence, as post-Soviet Russia discovered in the 2000s.

Additionally, ethanol burns more cleanly but less efficiently than gasoline. Harnessing efficient biomass fuel sources therefore means a reduction in the carbon emissions that produce global warming. Given the geopolitical import of climate change (including water shortages, population displacement, crop failures in traditional growing regions, the spread of parasites that thrive in warmer temperatures, an arctic region free of ice in summer and open to conflicts over mineral rights and territorial waters), both developed and developing countries have incentives to finance the development of at least comparatively green technology.

For these reasons, the race to develop biofuels is of tremendous strategic import. A wide variety of biomass material is under examination for potential fuel conversion, including garbage and cotton cloth (Purdue University, 2001b). DARPA has commissioned research into developing an affordable alternative to jet fuel made from farm animal feed to reduce military dependence on oil imports (DARPA, "BioFuels-Alternative Feedstocks," 2010). Other programs seek to engineer new crops that could be even more efficiently employed for this purpose (DARPA, "BioFuels," 2010).

Forward power projection also requires the availability of energy sources in the field, and biofuel presents military planners with the opportunity to follow the dictum of Sun Tzu that armies should consume the resources of occupied lands. DARPA is therefore funding research into mobile long-term reconnaissance robots that would scavenge biomass or garbage and then use it to produce their own fuel and to serve as electrical device charging stations for troops on forward deployments. Prototype EATR (energetically autonomous tactical robot) models ultimately to be used for reconnaissance were unveiled in 2009 (*Robotics Trends*, 2010). With DARPA-supported labs displaying free-running robots that could reach speeds of nearly fifty miles per hour in 2013 (Anthony, 2013), the addition of EATR technology to such devices portends armies that will not advance beyond their supply lines but instead will consume enemy resources as Sun Tzu advised in *The Art of War*.

Biomaterials and Biocomputing

Other avenues of research pursue the use of biomass for a variety of different purposes. The field of biological photovoltaics involves harvesting the energy production of plant photosynthesis using engineered proteins that would significantly boost the efficiency of solar energy devices. With a protein helmet coating supplying power to personal electronic devices, troops would have less to carry and would be able to operate in the field without the need to recharge equipment (Armstrong and Warner, 2003). "Protein-based photovoltaic coatings on Kevlar military helmets could produce enough energy for the soldier's electronics. Other equipment and vehicles could also be covered with these protein-based solar converters. A side benefit of such technology . . . is that the protein coatings would make whatever they coat more difficult to detect by electronic means since they would mimic the natural environment" (Purdue University, 2001b).

Biomaterials are "any material that is used to replace or restore function to a body tissue and is continuously or intermittently in contact with body fluids." Function-restoring examples include "kidney dialysis, prosthetic heart valves, hip replacement implants, and cardiac pacemakers." Priorities for development include new hemostatic bandages to treat field wounds but that cost just 10 percent of the price of the chitosan and fibrin-based Iraq War treatments that cost $1,000 per bandage (National Research Council, Committee on Capturing the Full Power of Biomaterials for Military Medical Needs [hereafter Committee, 2004]: 3, 11).

While new biomaterials (incorporating biological organisms or their outputs) must be reviewed and approved by the FDA for safety and efficacy, substances that are merely biomimetic (or "bioinspired") do not face this hurdle. One such example, developed by the US National Aeronautics and Space Administration (NASA) is the fastener Velcro. In addition to augmenting soldiers by giving them the proportional strength of insects, military planners also hope to endow them with lightweight body armor that absorbs impacts as efficiently as the exoskeletons of mollusks (Armstrong and Warner, 2003). "On a strength-to-weight basis, the abalone shell has armor protection capabilities equal to or greater than those of existing materials. . . . When laminated hierarchical structures of biological systems (e.g., the nacre of abalone shell) are mimicked . . . significant improvements in the composite mechanical properties have been observed" (Committee, 2001: 43).

The field of enzymology has also yielded significant advances in the development of resistant fibers and films. The development of new polymers permits the creation of seamless protective garments through the process of "electrospinning." Charging polymers with the correct voltage causes them to form jets of high-tensile fiber webs, with spaces of between .2 and 20 microns. Development began in 2000 of such a chem-bio protective-duty uniform, designed to be lightweight, launderable, and able to block infectious agents as well as flame and chemicals. The next step was enhancement with a selectively permeable membrane (US Army, *Analysis and Control of Polymer Interphases in Fibers and Films*, 1998), which is now commercially available for use in counterterrorism, decontamination, and force protection (W. L. Gore & Associates, 2014).

Other biological materials, such as a modified bacteriorhodopsin biomolecule that absorbs microwaves, might also be useful for avoiding radar detection if painted on tanks and aircraft (Committee, 2001: 44; Tian and Saaem, 2007). Bacteriorhodopsin, which changes its shape in response to stimulation by light, has been used commercially in other genetically modified forms to detect counterfeit currency, and it is also used in prototype holographic computer memory systems that store a quarter of the data that a typical personal computer can in a polymer vial no bigger than three centimeters across.

Advanced uniforms integrated with light wearable biocomputers that would support situationally aware and fully networked soldiers have been a goal of the US Army since the end of the Cold War (Armstrong and Warner, 2003; Committee, 2001: 27–28). In practice this has proven difficult to implement, with early twenty-first-century prototype programs culminating in the post–Iraq War Nett Warrior, a thirty-pound kit whose central components consist of commercially available smart-phone and tablet technology (Dixon and Henning, 2013).

However, subsequent developments demonstrated the capacity for development in biocomputing and the continuing interest of the military in R&D in this area. Researchers have discovered how to use a single gram of synthetic DNA to store seven hundred terabytes of data—or the equivalent of seventy thousand movies—indefinitely in a transportable freeze-dried form (Ingham, 2013). In 2015, the US Army Research Laboratory announced high-performance biocomputing as a goal to be achieved by 2030 (Trader, 2015).

Imaging and Surveillance

Enzyme research also entails the development of "bioreceptors" comprised of thin films with photoelectronic properties. Processes recently developed include integrating light-sensitive proteins into optical devices, particularly for laser eye protection, polymer-based batteries, and electromagnetic shielding. Bioreceptors can also detect the presence of selected DNA, which makes them useful in identifying infectious agents (US Army, *New Materials Development Using Biotechnology Process*, 1998). Although, as described in the next chapter, biosensors are currently not capable of providing useful real-time data for homeland security monitoring, in the future, "a network of biosensors could considerably improve a commander's view of the battlefield. Some researchers envision soldiers wearing wristwatch-style biosensors that are sensitive to a variety of target molecules. In effect, each soldier would become a detection device and warn of a possible biological or chemical attack. Also, such sensors could be used to monitor the health and wellbeing of entire units" (Armstrong and Warner, 2003).

Other biotechnologies would grant increased surveillance capabilities to individuals involved in defense and national security. The artificial retinas given to troops with damaged eyesight could also be used for high-resolution imaging (Committee, 2001: 29). Other biotech efforts are intended to permit more rapid and efficient processing of collected data: "Successful development of a neurobiologically based image triage system will increase the speed and accuracy of image analysis in a context where the number of acquired images is expected to rise significantly" (DARPA, "Neurotechnology for Intelligence Analysts," 2010).

But still other efforts have made tremendous progress in reducing the role of humans in collecting data and replacing them with other agents: insects. Although the US Army may have experimented with mosquitoes as bioweapon delivery systems in the 1960s (Maurer, in Maurer, 2009: 96), in 2009 DARPA-funded engineers at the University of California, Berkeley, announced that they had developed cyborg beetles that they could direct by remote control. The researchers implanted electrodes into the brains and muscles of two species of beetle, which could then be made to fly and maneuver on command. "The project's goal is to create fully remote-controlled insects able to perform tasks such as looking for survivors after a disaster, or acting as the ultimate spy" (Callaway, 2009). DARPA describes its Hybrid Insect MEMS Program as follows:

> The HI-MEMS program is aimed at developing tightly coupled machine-insect interfaces by placing micro-mechanical systems inside the insects during the early stages of metamorphosis. . . . The intimate control of insects with embedded microsystems will enable insect cyborgs, which could carry one or more sensors, such as a microphone or a gas sensor, to relay back information gathered from the target destination.
>
> HI-MEMS derived technologies will enable many robotic capabilities at low cost, impacting the development of future autonomous defense systems. The realization of cyborgs with most of the machine component inside the insect body will provide stealthy robots that use muscle actuators which have been developed over millions of years of evolution. (DARPA, "HI-MEMS," 2010)

Prior to the insect agents, DARPA had already created a "roborat," a rodent controlled by a neural prosthesis via a laptop keyboard so that it could climb stairs and navigate mazes, which Director Goldblatt compared to a child's remote-controlled car. Further developments included mounted cameras for visual data collection and the prosthesis implanted along the rat's belly so that it would not be observed. The neural prosthesis stimulated the rats' pleasure centers, motivating them in their tasks, and Moreno noted that there are obvious implications for how such prostheses could be adapted to human subjects as well (Moreno, 2006: 43–44).

Within a decade, in 2015 DARPA launched its new Biotechnology Program Office (BTO) with a two-day conference headlined by DARPA head Arati Prabhakar. Among the journalists DARPA invited was a representative of *Humanity+* magazine, a transhumanist journal. In its favorable coverage of the event, *Humanity+* proclaimed that "DARPA's new vision is simply to revolutionize the human situation and it is fully transhumanist in its approach." In particular, the article devoted considerable attention to BTO lead researcher Geoffrey Ling's presentation of DARPA-supported work on a "cortical modem . . . a direct neural interface that will allow for the visual display of information without the use of glasses or goggles."

> First, this technology could be used to restore sensory function to individuals who simply can't be treated with current approaches. Second, the device could replace all virtual reality and augmented reality displays. Bypassing the visual sensory system entirely, a cortical modem can directly display into the visual cortex enabling a sort of virtual overlay on the real world. Moreover, the optogenetics approach allows both reading and writing of information. So we can imagine at least a device in which virtual objects appear well integrated into our perceived world. Beyond this, a working cortical modem would enable electronic telepathy and telekinesis. The cortical modem is a real world version of the science fiction neural interfaces envisioned by writers such as William Gibson. (Rothman, 2015)

EXOTIC WEAPONRY

Whether with a host of cybernetic insect spies or a company of super-soldier Augments with the abilities of insects, the United States and its technologically advanced allies and competitors are assuming the capacity to wage conventional warfare and espionage in a manner that will not soon be available to internal or regional adversaries or to non-state antagonists. But it is in the area of novel bioweapons where hegemonic actors stand poised to offer attacks against which their adversaries could mount no possible defense. Currently many potential lines of research are banned under the terms of the BWC, but even if state actors abide by its terms, private-sector breakthroughs will continue to have dual-use capabilities that can be studied. Indeed, some of them have already caused outbursts of political violence internationally.

Genetic Weapons

Until the end of the twentieth century, bioweapons meant pathogens (and possibly animal delivery systems). The biotech revolution, particularly the ability to sequence and translate entire genomes, has altered that equation. Some state militaries, notably China's, are already publicly expressing an interest in attacking targets by reordering their bodily functions through what is known in more benign applications as gene therapy. Planners in the United States also note that "the long term implications of genomics will present the Army with opportunities and challenges even in the next decade. . . . The Army can, however, promote development of new products and processes that will be consistent with or specific to its missions and needs. This will require that the Army be fully aware of the synergistic effects of biological tools" (Committee, 2001: 15).

"The goal of gene therapy is to effect a change in the genetic makeup of an individual by introducing new information designed to replace or repair a faulty gene." This is accomplished by using the same principle employed since the first smallpox vaccination: the use of a harnessed, crippled virus to serve as a "Trojan horse" vector, in this case bearing replacement or supplemental genes to alter cell functioning. Somatic cell therapy affects only the cells of the individual receiving it, and for reasons of bioethics and technical feasibility, most therapeutic research has been of this type. But there is also the technique of germline cell therapy, used in the GloFish described in the last chapter, which might "lead to a heritable change that could repair problems for all future generations" (Block, in Drell, Sofaer, and Wilson, 1999: 60, 62).[8] Needless to say, while it is appealing to imagine a future in which genetic disorders have been eradicated, it is just as easy to imagine that the technology could be used to cause intentional harm as well, just as the British used their superior understanding of smallpox and access to samples to the detriment of the Delaware Indians.

Although American military planners are bullish on the potential for gene therapy to improve the lots of wounded servicemen in the near future, the technologies are not yet universally acclaimed nor even accepted. As of 2015, the FDA had approved some cellular therapies but had not yet approved any human gene therapy products for sale (US FDA, 2015). Current gene therapy is experimental and has not proven very successful in clinical trials. Little progress has been made since the first gene therapy clinical trial began in 1990." This reaction stems in part from the death and illness of several children who received gene therapies to treat life-threatening chronic conditions. At the same time, however, researchers elsewhere announced that gene therapy safely and successfully restored partial sight to congenitally blind test subjects. The results were accomplished by inserting healthy copies of a missing gene into patient retina cells via a vector manufactured by a private American company called Targeted Genetics (University College of London, 2008).

Vector-delivered gene therapies remain an emerging biotechnology, but cases such as these demonstrate both that vectors can be used to create significant physical alterations in targets and that these changes can be deadly. The discovery that viruses can be carried airborne for considerable distances even after the droplets of fluid constituting their transmission media have fallen to the ground provides further evidence that vectors might soon be used to deliver genetic therapies—or maladies—to wide target populations (Medical News, 2007). With the genetic maps of entire organisms now available—the full genome for the plague bacterium was decoded in 2001— it is inevitable that researchers will develop the means to rewrite specified segments of targeted genes (Preston, 2009: 296).

Direct-Effect Weapons

The US military is currently developing "a set of design and synthesis processes that will enable the specification of a desired function, and be able to rapidly synthesize a protein that performs the function." Rather than modifying existing proteins, this biotechnology would allow the creation of new proteins based on specific performance objectives (DARPA, "Protein Design Processes," 1998). The field of genetic protein decoding and engineering of this kind is known as proteomics (Committee, 2001: 15).

Understanding the functions of proteins is key to opening entirely new frontiers in medicine—and warfare. Already, researchers have destroyed targeted cancer cells by using engineered nanoparticles to deliver genes only to the tumor and not to healthy neighboring tissue. Once the genes were inserted, they stimulated the production of a protein that selectively destroys the cancer (BBC News, 2009).

Along with the promise of proteomics comes the peril. Infectious protein particles, discovered in 1982, are called prions. Like bacteria and viruses, prions are transmissible and cause damage to organisms by converting existing proteins into prions, leading to more infection and preventing the organism's cells from performing their normal functions. But prions are not susceptible to any known remedies. Because they are currently untreatable, they are always fatal, as evidenced by human and animal victims of Creutzfeldt-Jakob disease and bovine spongiform encephalopathy (BSE), or mad cow disease.

Prions, a misfolded form of natural proteins, could be created using modified recombinant DNA or by using peptide synthesizers "widely found throughout research and medical communities." Obviously the surreptitious introduction of BSE into cattle stocks would cause tremendous damage to both industry and public health. Even uncorroborated reports of mad cow disease are enough to cost millions of dollars in lost sales to agribusinesses (Preston, 2009: 296–297, 308). Synthesized prions truly offer the old Biopreparat vision of a highly transmissible and incurable bioweapon.[9]

A different avenue of potential development in biotechnological attacks is a shift away from infectious agents to targeting human bioregulators, natural substances in the body that control automatic processes such as blood pressure and immune responses. Alibek claimed that the Soviet Union pursued this research into "direct-effect weapons" in the 1980s to circumvent the BWC. As described in chapter 2, the result would not actually be an illness, but the turning of the body against itself through disruption (Preston, 2009: 313–314).[10]

Chinese researchers Guo Ji-wei and Xue-sen Yang (2005) directly addressed the security applications of such efforts in proteomics, arguing that

> direct-effect weapons . . . can cause destruction that is both more powerful and more civilized than that caused by conventional killing methods like gunpowder or nuclear weapons. . . . A military attack, therefore, might wound an enemy's genes, proteins, cells, tissues, and organs, causing more damage than conventional weapons could. However, such devastating, nonlethal effects will require us to pacify the enemy through postwar reconstruction efforts and hatred control. . . . [W]e could create a microbullet out of a 1 micron tungsten or gold ion, on whose surface plasmid DNA or naked DNA could be precipitated, and deliver the bullet via a gunpowder explosion, electron transmission, or high-pressured gas to penetrate the body surface. We could then release DNA molecules to integrate with the host's cells through blood circulation and cause disease or injury by controlling genes.

Around the same time, an American biodefense expert added that "if one can disrupt unit loyalty through fear or another emotion, the US Army would cease to exist as a fighting force. Claustrophobia would make soldiers tear off their protective face mask. Fear, thirst, accelerated heart rate, hypermotility of the gut—these would be the desired peptide effects." Delivery would be accomplished using engineered pathogens, and their primary role in biowarfare would be as delivery systems for direct-effect weapons rather than the transmission of infectious disease (Moreno, 2006: 178–179). Indian military analysts describe direct-effect weapons as "more civilized than conventional killing methods," but also "likely to reveal a character of aggression not thought of to date" (Singh, 2008).

A variety of emerging biotechnologies provide the foundational research for such direct-effect weapons. Beyond the study of individual proteins in proteomics, metabolomics examines how they work in concert to produce the body's metabolic systems, toxicogenomics pinpoints causes of chemical toxicity, and pharmacogenomics explores the genetic basis of drug failure and side-effects in certain people. "Microarrays the size of a postage stamp can analyze your entire genome in days. . . . Imaging devices visualize single molecules inside cells so their activities and chemical facets can be studied" (Klotz and Sylvester, 2009: 18).

Future Avenues of Biowarfare

The ability to obtain detailed information about every bodily function at the biochemical level invites cuts with the same double-edged sword as other biotech developments have. The possibility of healing and strengthening entire populations with access to these resources coexists with the recognition that the salutary use of biotechnology depends entirely on the will of the wielder. Technologically advanced states will be the only actors with the capacity to develop novel direct-effect weapons in the foreseeable future and also have the opportunity to expand their economic dominance. Even if they do not avail themselves of genetic arsenals, they will be in a position to use the technologies to enhance the physical well-being of their own populations, potentially increasing the productivity of their workforces while lowering conventional health-care costs.

DARPA's Narrative Networks (N2) program examined brain activities such as the neurological processes occurring when subjects felt empathy, for data about how to use stories to "facilitate faster and better communication of information in foreign information operations" (Miranda et al., 2015: 62). Critics contended that the military wanted to create propaganda software that could continuously update its spin power to persuade audiences in any manner necessary (Dewar, 2015). Given the prior work of project leader William Casebeer on counterterrorism and the hand-wringing during the 2010s over the apparent inability of security agencies to match what was described as the compelling narrative spun by the Islamic State of Iraq and Syria (ISIS) to recruit thousands of Westerners, the rationale and appeal of N2 are evident. Comic book writer Grant Morrison (2011: 415), in discussing Casebeer's work on "counter-narrative strategies," finds the implications of narratives "as addictive as cocaine" lies not just in discrediting enemy rationales for fighting but in the capacity "to develop a technology whereby a cadet is told a story so convincing he believes he's superhuman before a battle."

Some biotechnologies that could easily be used as instruments of political control are already available. Researchers evaluating patients who suffer from a rare genetic disorder called Williams syndrome that renders victims "pathologically trusting" have determined that their brains produce too much of the hormone oxytocin, which they have dubbed "the trust hormone."

As one researcher, Paul Zak, described it, "if you just had high levels of oxytocin, you would be giving away resources to every stranger on the street."

> In 2001, Zak began spraying oxytocin up the noses of college students to see if the hormone would change the way they interacted with strangers. It did. Squirt oxytocin up the nose of a college kid, and he's 80 percent more likely to distribute his own money to perfect strangers. This gave Zak an idea. Like some comic-book villain concocting a plan to take over the world by dumping

happy pills in the water supply, he wondered if it might be possible to use this molecule—oxytocin—to change the way people felt about the government. . . . Zak put 130 test subjects through his normal routines. He sprayed half of them with oxytocin, half with a placebo, then ran them through a battery of tests and measurements. "The people on oxytocin did report that they trusted other people more, and the people who trusted others more also trusted their government more." (Spiegel, 2010)

One private company that sells pheromones, biochemicals used in communications by insects and marketed as human aphrodisiacs, is already selling oxytocin spray as Liquid Trust (Klotz and Sylvester, 2009: 34). It is easy to imagine a wide range of potential applications for manipulating targets biochemically, not least through pacification efforts directed at belligerents or rioting civilians. It is not clear that spraying oxytocin around foreign civilian populations in the vicinity of military installations would constitute a violation of the letter of the BWC.

Other possibilities for advanced biowarfare include returning to classical bioweapon approaches, but employing engineered "designer" bacterial and viral agents. This could involve rendering bacteria resistant to particular antibiotics, a possibility that both the United States and Soviet/Russian programs have explored, but also using DNA shuffling (polymerase chain reactions or PCR, described in the previous chapter) to speed mutation up to thirty-two thousand times faster than what occurs through natural selection, meaning that bacteria would develop their inevitable resistance to pharmaceuticals very rapidly (Block, in Drell, Sofaer, and Wilson, 1999: 57–58).

Characteristic rapid mutations common to RNA viruses make them difficult to counter, and the mutagenic drift of the H1N1 virus (swine flu) produced a global epidemic of a pathogen that otherwise was unfamiliar to most humans in 2009. Another example of mutagenic drift can be found with parvovirus, a common threat to domestic dogs, which had been observed only as feline distemper before the 1970s and which has more recently produced strains that have infected humans. Such transgenic shifts are rare, but genetic engineers have the power to produce host-swapping diseases (Block, in Drell, Sofaer, and Wilson, 1999: 65–68). Indeed, gene therapy technology demonstrates that "engineers unable to obtain exotic pathogens might be able to manipulate common vectors to produce the desired effects," although the risk of genetic drift is obvious (Preston, 2009: 301, 207).

Designer diseases might also include "stealth viruses" that infect human cells but remain indefinitely inactive until triggered by a controlled signal. Block (in Drell, Sofaer, and Wilson, 1999: 63) claims that "stealth viruses could be designed to be contagious, and therefore distribute themselves silently throughout a given population. They might even be designed against specific target groups."

Such thinking inevitably leads to the question of whether it is feasible to create bioweapons that will selectively target particular groups, such as pathogens that activate only in the presence of particular genetic markers. Saddam Hussein publicly claimed to be interested in developing weapons that would target Jews, and an unsubstantiated story in *The Times* of London in 1999 claimed that Israel was working on an "ethnic bomb" in response. The story also cited a source from the British Chem-Bio facility at Porton Down as describing such weapons as theoretically possible (Block, in Drell, Sofaer, and Wilson, 1999: 47–48), but subsequent reports have dismissed the story, and—for now—the possibility of an ethnic bomb as a hoax (Preston, 2009: 296). At the same time, Alibek (1999: 268) argued cryptically that Russia "has every interest in bioweapons for contemporary problems" (which were primarily Chechen separatists at the time), and Moscow banned the export of human biological samples due to concerns that they could be used to develop ethnic bioweapons (Klotz and Sylvester, 2009: 28).

Finally, another possibility for gene therapy to be used as a weapon is in the area of programmed cell death, or apoptosis. All cells are naturally programmed to stop growing at particular equilibrium points; those that do not stop become cancerous tumors. The healthy development of all living organisms requires effective cell life cycles, and many human cancer patients have been observed to have missing or mutated tumor-suppressing genes. However, the ability to initiate cellular apoptosis by rewriting genetic codes would provide a terrible destructive capacity: if researchers deployed a vector that "activated death pathways in all cells, it would potentially be more toxic than any known poison!" (Block, in Drell, Sofaer, and Wilson, 1999: 68–70).

The Significant Footprint of the Private Sector

Engineered bacteria or viruses would need to be easily eliminated by their creators to prevent epidemics. One solution with far wider implications is the use of a "terminator" gene that would either ensure that organisms are incapable of reproduction, or a type that would destroy its own cells if exposed to certain enzymes. Rather than a speculative technology, terminators are a reality that has been causing international tensions since their commercial introduction in the mid-1990s. American seed companies with a significant export market to developing countries, among them, Delta and Pine Land Inc., incorporated terminators into wheat and rice, ensuring sterility after the first yield and the necessity of repurchasing seed annually. After the introduction of terminator crops in rural India led to riots and attacks against foreign holdings by irate farmers, the Indian government was spurred to invest in the development of India's own biotech industry (*Asian Age*, 1998).

In 2007, Delta and Pine Land was purchased by Monsanto, the herbicide manufacturer that previously had provided the highly carcinogenic defoliant known as Agent Orange employed by the United States in Vietnam. Monsanto is also one of the agribusinesses credited with the "green revolution" of the twentieth century, in which imported and specially bred plants were introduced to developing countries to increase crop yields. This development is credited (or blamed) for increasing the world population by at least one billion human beings. The corporations responsible for this development argued, as pharmaceutical companies would later, that it was necessary to protect intellectual property rights if biotech development were to continue, and in 1980 the US Supreme Court ruled that private companies could patent engineered biological organisms. The result was a progression to terminator-equipped seeds that produce 5 to 10 percent greater crop yields and more nutrients thanks to the introduction of algae genes, as well as greater resistance to drought. Monsanto GMOs are now also resistant to the company's signature herbicide, Roundup. But farmers who availed themselves of the products were required to contractually accept that the GMOs contain terminators and that they would be required to buy new seed annually or be permanently barred from purchasing the enhanced seeds (*The Economist*, November 21, 2009). In the 2000s, the UN voted repeatedly for a moratorium on commercial use of terminators (Todhunter, 2013).

Klotz and Sylvester (2009: 14) note the possibility of corporations attempting to "undermine a rival with pathogens," and agribusiness would seem the most obvious sector in which such attacks might occur. But the consequences to states would also be severe. An outbreak of foot-and-mouth disease (FMD) decimated the British agriculture and tourist industries in 2001, with losses to the beef industry alone coming in at $17 billion. In 2004, the importation of a single pig from Hong Kong infected with FMD caused the loss of $19 billion to the Taiwanese pork industry. The US Department of Agriculture reports that repeated avian flu outbreaks cost the American poultry industry hundreds of millions of dollars, and it estimates that a single effective biological attack could cost $10 billion to $30 billion to the nation's agricultural sector (Preston, 2009: 317–318; Segarra, 2002).

Again, however, the dual-use potential of many biotechnologies means that the easy destruction of crops and livestock does not need to remain the province of multinational corporations. The military implications of terminator genes and vectors bearing agricultural pathogens are obvious. The real question is whether or not states are willing to target food supplies through the release of engineered vectors that would cross-pollinate targeted crops with terminator genes. Such a strategy could be viewed as a form of direct sanctions more humane than conventional warfare or, alternatively, as unjust collective punishment.

The private research sector is also making other discoveries that have potentially profound implications not only for warfare but also for the future of the human species. In 1998, the Geron Corporation, located in California, was at the forefront of these advancements with the announcement of accomplishments including cultured human stem cells that could ultimately be used for organ regeneration. Less noticed was that the company also announced that the cause of aging in living cells had been identified and could be slowed to half the rate of the natural process. The telomerase enzyme, produced by the telomeres at the end of chromosomes whose exhaustion signals the end of cell division, can be added to permit the rejuvenation of damaged and diseased cells. These short-term implications obviously carry significant implications for the rapid treatment and rehabilitation of wounded troops (Wade, 1998).

Three months later, Geron carried this research one step further by announcing that experiments demonstrated that the absence of telomerase, abbreviated TR, causes tissue and organ dysfunction. Telomerase-deficient mice suffered what was whimsically dubbed a "TRKO," characterized by progressive defects in the immune and reproductive systems (Geron Press, 1998). The purpose of the experiment was to demonstrate that telomerase could be used to effectively treat age-related diseases and cancer. Yet it is quite easy to conceive of a DNA virus that is programmed to deactivate the genes that produce telomeres. With the advent of a virus that could produce rapid telomere destruction, an "aging bomb" seems a distinct possibility.

These discoveries also accelerated research on "immortalization." While living forever remains out of reach, those who can afford genetic therapy will have the opportunity to lead longer, more productive lives than those who cannot (Associated Press, 1998). And, as Egudo (2004: 15) notes, developments from such technologies can be used to keep more of the population in condition for military service at more advanced ages, an important consideration for industrial states with graying populations.

THE INTERNATIONAL BALANCE OF POWER

With the emergence of advanced biotechnologies, many of which already exist or are being developed for expressly military purposes, the United States holds the potential for achieving a decisive advantage in power projection capabilities beyond the reach of its current adversaries and most of its likely potential competitors. Besides the United States, other actors are expanding their biotech R&D sectors. "As the Chinese military expands its power projection capabilities, it will concentrate on creating asymmetrical advantages in the face of superior US conventional technology" (NTI, 2003).

India, with its reliance on the green revolution to attempt to achieve food sufficiency, has spent the last two decades encouraging the development of agricultural biotechnologies.

Many of these advances were facilitated using extensive knowledge of genetic engineering, which in turn provided information on the de novo synthesis of biological agents. Whether such synthesis has actually been done is uncertain. India has made substantial efforts to prepare its military force for a biological attack. In December of 1998, India began to train its medical personnel to deal with the eventualities of such an attack. (NTI, 2009)

Today, India's equivalent of DARPA, the Defense Research & Development Organization (DRDO) operates a network of fifty-two laboratories whose research includes life sciences for military purposes. These include the Defense Institute of High Altitude Research and the Defense Food Research Laboratory (Department of Biotechnology, Government of India, 2013: 20). Its reported products parallel those investigated by its American counterparts, including treatments to combat altitude sickness, transgenic crops, and protective polymers for uniforms, although products are frequently described in terms of their commercial rather than strategic potentials (Defense Research & Development Organization, 2015).

India's biotech experience is hardly unique. The dual-use dilemma exacerbates security dilemmas, situations also caused when actors interpret defensive measures as preludes to offensive assaults. Thucydides recorded one significant instance in describing the fortifying of Athens, and the Cuban missile crisis represented another that nearly resulted in an apocalyptic nuclear exchange. But military biotech R&D is not always meant to produce weapons. Alibek noted that the United States' Pine Bluff Arsenal, developed for bacteriological weapons production, had switched to research on immunosuppressive substances for organ transplants (Alibek, 1999: 237). Likewise, the US Army has conducted clinical trials of a multistrain HIV vaccine in Thailand (Marble, 2010).

Even if the great powers of the twenty-first century do turn their biotech sectors toward an arms race as they did in the previous century, they are still subject to deterrence through the same threats of massive retaliation issued at the height of the Cold War. One possible response by rogue states could be the clandestine transfer of CBRN material to non-state actors who would then act on their behalf, a concern cited as significant enough to justify preemptive war against Iraq and continued engagement with flawed regimes in Pakistan. The underlying assumptions behind this threat are that terrorists want CBRN weapons and that sympathetic states would be willing to share them, either in support of their cause or so that non-state actors would be blamed for attacks masterminded by governments that could maintain plausible deniability. However, this presumes that authoritarian regimes would trust actors outside of their direct control and hierarchy with sensitive material, and furthermore trust them to follow their established foreign policy objectives. This strategy would leave such rogue states probably more vulnerable than empowered, and thus they are unlikely to proliferate to non-state actors (Whitlark and Stepak, 2010).

Perhaps the greatest threat to international stability in the genomic age is the international emergence of two classes of humankind separated by disparities in living conditions far wider than those between the developed and developing worlds today. Described by biologist Lee Silver, this would be "a two-class system with rich, genetically enhanced 'GenRich' types lording it over poorer, inferior 'Naturals'" on a global scale (Armstrong and Warner, 2003).[11] Ultimately, the perception of injustice by the multitude of the have-nots would render such a system unstable (Carr, 1939). While matching advanced technology is a challenge to would-be competitors, it is not an insurmountable one (Quille, in Lewer, 2002: 45). And it might actually inspire new forms of lower-cost asymmetric counterattacks, as attempts to use model airplanes as drones to attack American targets by would-be terrorists demonstrates.

But for now, as in the nuclear club, with their overwhelming edge in both offensive and defensive capabilities, the United States and other advanced industrial nations can rest assured that their military and economic dominance of the international system is in no jeopardy. Biotechnology, often cited as an asymmetric threat to conventional power projection capabilities, is being harnessed by those very militaries as a force multiplier, and their R&D and production capabilities far outstrip those of any possible combination of rogue states and terrorist groups. The biotech RMA is well underway, and states are free to shift their attention from international to internal threats.

NOTES

1. Coincidentally or not, this was precisely the depiction of Darth Maul five years later in *The Phantom Menace*.

2. I borrow the term *Augments* from a 2004 storyline in the television series *Star Trek: Enterprise* that depicted genetically enhanced super soldiers. This squad of fighters was genetically modified for speed and strength as embryos rather than being equipped with other technologies during the course of military service.

3. According to the US Army surgeon general, nearly 8 percent of the army was on sedatives and more than 6 percent on antidepressants, or more than 110,000 personnel, in 2011, figures that were comparable to their use by the civilian population (Murphy, 2012).

4. Genetically modified mice have demonstrated a greater capacity for rapid learning (Schreiweis et al., 2014).

5. One such result is the Legged Squad Support System (LS3), an all-terrain robot, described variously as a pack mule and an Imperial walker from the *Star Wars* films, which can conduct all-terrain surveillance or one-way bombing attacks (Michael, 2012).

6. One stated response of the Saudi government—to shift its efforts away from oil and toward other industries in which it has a competitive advantage, such as dried date exports—hardly seem sufficient to avert fundamental geopolitical shifts (interview with Saudi embassy official, Washington, D.C., September 18, 2008).

7. In 2006, President George W. Bush declared in his State of the Union address that the United States had "an addiction to oil" and urged the development of biofuel from switchgrass.

8. Technically, however, the institution of heritable germline changes in a species, including humans, produces a new species or subspecies (Juengst, in Savulescu and Bostrom, 2009: 52).

9. For a fictional depiction of engineered prions that produce weaponized Marburg, leaked from Biopreparat and eventually into the hands of an apocalyptic sect, see the 1998 episodes of the television series *Millennium*, "The Fourth Horseman" and "The Time Is Now."

10. Huang and Kosal (2008) report that the US Air Force has considered and rejected one type of bioregulator attack: the development of a neuropharmaceutical aphrodisiac, to be sprayed over enemy ground forces, intended to function as a "gay bomb."

11. Perhaps the bleakest expression of this possibility, redolent of *Star Trek* archvillain Khan, comes from George Annas's 2000 article "The Man in the Moon": "Ultimately it almost seems inevitable that genetic engineering would move homo sapiens into two separable species: the standard-issue human beings would be seen by the new, genetically enhanced neohumans as heathens who can properly be slaughtered and subjugated. It is this genocidal potential that makes species-altering genetic engineering a potential weapon of mass destruction and the unaccountable genetic engineer a potential bioterrorist" (Juengst, in Savulescu and Bostrom, 2009: 48).

Chapter Four

Homeland Security, Human Security

While the US military allocated billions of dollars in the first decade of the twenty-first century toward advanced BW defense and augmentation of individual warfighters, an equally dramatic expansion of nominally civilian R& D was also occurring under the rubric of homeland security. Although it had begun in the 1990s, it was propelled by the 9/11 and Amerithrax attacks in the fall of 2001, which clearly demonstrated the vulnerability of major industrialized states to low-cost asymmetric attacks within their own borders. They also indicated how infrastructure, such as transportation systems and the US Postal Service, could be exploited as conduits of terrorism.

Responses to the newer and deadlier forms of terrorism built upon the foundations of the previous century's BW programs, taking advantage of the dual offensive/defensive nature of the research involved. But while the industrial-scale BW programs of the twentieth century were products of highly centralized state military programs, much of the current homeland security biodefense work is conducted at least partially in the private and academic sectors without full governmental oversight. This is consequential because, dating back to the inoculation programs of the British military in the 1700s, the development of defenses against biological agents requires work with the production and refinement of specimens of the very same pathogens against which defenses are being developed. In the past, technologies developed for civilian medical or agricultural purposes have been easily adapted to the production of deadly infectious agents or the engineering of vectors to create entirely new weaponized diseases.

As the United States and other advanced industrial nations develop new military applications of biotechnology to ensure their continued control of international security, they have also moved to prevent threats to domestic security. While state BW programs were originally intended to terrorize civilian populations and disrupt agriculture and public transportation systems, such attacks are currently expected to come from non-state actors rather than state-sponsored fifth columnists. Transnational entities such as the Takfiri Islamists of al Qaeda or doomsday cults such as Aum Shinrikyo have demonstrated both the willingness and the ability to perpetrate mass-casualty attacks. What were apparently purely domestic non-state actors, including the perpetrator of the Amerithrax attacks and a number of right-wing militias, have also been caught attempting to obtain or have actually acquired rudimentary biological and chemical arsenals.

Although terrorism is described by analysts as a "weapon of the weak," the specter of mass-casualty CBRN attacks by these non-state actors is frequently cited by political leaders as the gravest current threat to national security. For this reason, the US Congress passed the BioShield Act of 2004 to provide a stockpile of prophylactics and treatments, increasing biological defense—and, unavoidably, offense—research expenditures by over $50 billion (Klotz and Sylvester, 2009: 1–7, 98). A great deal of controversy quickly surrounded the program, both in terms of its likely effectiveness and the lack of oversight and security attached to sensitive materials.

Other ethical questions attached to these homeland security efforts, including the equitable distribution of safeguards and the appropriate protection of individual health information privacy, have also arisen with the new biodefense programs or else have remained unanswered since the Amerithrax attack led to their establishment. Just as the advent of a biotech military revolution to eliminate vulnerability on the battlefield portends both promise and peril, efforts to eliminate domestic vulnerability through the establishment of a biodefense-industrial complex also presents its own challenges that must be weighed against the threat of attack.

THE THREAT OF TERRORISM

An accurate probability of the occurrence of significant bioterrorism incidents is likely unknowable. National security planners began to predict that terrorists would soon acquire CBRN weapons as long ago as World War II. Despite these dire warnings, and the fact that the superpowers were rapidly assembling such arsenals, there have been very few instances in which non-state actors have actually been observed attempting to deploy biological or other weapons of mass destruction (Maurer, in Maurer, 2009: 1).

In 1946, Dahm Y'Israel Nokeam (Avenging Israel's Blood, or DIN), a hard-line faction comprising seven members of a wartime Jewish partisan fighting group, decided to continue its retaliation for the Holocaust. After dismissing several plans for mass contaminations because they would inevitably afflict the American occupation force, DIN ultimately succeeded in poisoning the bread of Nazi prisoners at the Stalag 13 POW camp at Nuremberg. Official records indicate nearly 2,500 casualties, with the number of fatalities unreported, while DIN claimed that 4,300 became ill and several hundred more dead or paralyzed. The faction ultimately escaped to Israel, but some members returned in 1948 with an ultimately unsuccessful plan to poison the water supply of the civilian population with an arsenic compound (Sprinzak and Zertal, in Tucker, 2000: 17–40).

In 1972, a Chicago ecoterrorist group that wanted to eliminate most of the human race were arrested by police as they were preparing to dump typhoid bacteria into the city water supply. The cultures were produced by a student who had not even completed one community college course in microbiology but had an internship working in the University of Illinois Hospital, from which he tried to obtain samples of several pathogens that he intended to employ simultaneously to overwhelm biodefense measures (Carus, in Tucker, 2000: 55–67).

In 1981, a group calling itself the "Dark Harvest Commando of the Scottish Citizen Army" took anthrax-contaminated soil from Gruinard Island, a World War II BW testing range left contaminated by the British Army that the groups demanded be remediated, and placed it outside the BW facility at Porton Down in England. It also left uncontaminated soil from the same site outside a Conservative Party retreat in Blackpool and reportedly planned in late 2001 to use additional anthrax-contaminated soil against Prince William, to whom it had already sent innocuous white powder. The group has also threatened to poison water supplies in London and elsewhere in England if British troops are not removed from Northern Ireland (Committee, 2006: 44; *The Scotsman*, 2002).

In January 2014, months before ISIS declared a caliphate, its fighters fled a battle against a Syrian rebel group and left behind a computer that came to be known as the "Laptop of Doom." The laptop's owner had been "a Tunisian national named Muhammed S. who joined ISIS in Syria and who studied chemistry and physics at two universities in Tunisia's northeast. Even more disturbing is how he planned to use that education: The ISIS laptop contains a 19-page document in Arabic on how to develop biological weapons and how to weaponize the bubonic plague from infected animals." It was accompanied by a fatwa from an imprisoned Saudi cleric that justified BW attacks against the infidels as a last-ditch measure, "even if it kills all of them and wipes them and their descendants off the face of the Earth." When contacted by reporters, his alma mater claimed to have lost track of him three years earlier, but its representative asked unprompted, "Did you find his papers in Syria?" and deferred further questions to Tunisian state security (Doornboss and Moussa, 2014).

A 1994 study that included the deliberate contamination of food, water, and drugs in its definition identified more than 244 incidents of terrorism using chemical or biological weapons in twenty-six countries since World War I, with 60 percent involving the actual use of the agent and another 10 percent involving the acquisition of it, rather than just a threat. Most of these were classified as criminal acts, with only one-quarter ascribed political motivations (Tucker, 2000: 1).

Beyond the difficulties in obtaining supplies of BW, there is the question of whether a non-state proliferation demand actually exists. Whitlark and Stepak (2010) coded the historical record of thousands of terrorist and insurgent attacks over the previous forty years to determine patterns of CBRN usage. Their study yielded the finding that, despite the ready availability of biological material that could be employed as weapons—as the Assyrians and Romans could attest—biotechnology has only been incorporated into a handful of attacks. A National Defense University study indicated that in the twentieth century there were nineteen instances in which non-state actors employed biological agents, with most involving criminal activity rather than political terrorism (Kortepeter and Parker, 1999: 524). Ball and colleagues (in Maurer, 2009: 487) note that "despite recurring fears, actual historical development and use of biological weapons has been limited."

Preston argues that it is extremely unlikely that an isolated "kook in the basement" could successfully pull off a large-scale bioterror attack, but groups could be successful if they enlist even "one competent microbiologist" with a graduate education and a few thousand dollars. In fact, a US government project codenamed Bacchus demonstrated in 1999 that a simulated terrorist group was able to assemble for $1.6 million a "mini-biological weapons plant using off the shelf technology ordered from commercial outlets that produced two pounds of simulated anthrax" (Preston, 2009: 188).

However, when anthrax attacks did occur in the United States in late 2001, the culprit was apparently a lone actor, military research scientist Bruce Ivins, who had access to highly virulent anthrax samples, the knowledge and skill to concentrate them into a highly dispersible fine powder, and the motivation to sustain flagging research funding by attracting attention to the potential threat. The Amerithrax case resulted in twenty-two individual anthrax infections (half of which were inhalation and half cutaneous), as well as an estimated $250 million in direct decontamination costs, primarily of the Hart Senate Office Building where one of the envelopes was opened in the office of Senate Majority Leader Tom Daschle.[1] Another envelope filled with refined *B. anthracis* spores was sent to Senate Judiciary Committee chairman Patrick Leahy but was not discovered until days after the Daschle letter had been opened (Koblentz, 2009: 204–205).

These were not the only biological attacks against the federal government during this period, as the deadly substance ricin (made from castor bean extract) was discovered in the White House mail room in 2003 and the office of Senate Majority Leader Bill Frist the following year (Koblentz, 2009:

204–205). Ricin had also figured into a 1994 plot by four antigovernment activists who were the first people to be convicted and sentenced to prison terms under the 1989 Biological Weapons Anti-Terrorism Act. These members of the Minnesota Patriots Council, who blamed the ruinously low farm prices of the 1980s on a plot by the US Federal Reserve and "Jewish bankers" to drive them out of business and purchase their land, planned to use ricin to assassinate local and federal law enforcement officials. They spent twelve dollars to obtain castor beans and instructions for eliminating "evildoers" through the mail via an advertisement in a right-wing publication placed by an extremist group in Oregon selling "Silent Tool of Justice" kits (Tucker and Pate, in Tucker, 2000: 159–168).

And, in two separate but nearly simultaneous cases in 2013, two different individuals, both with show business careers and connections to antigovernment activism, and who both attempted to frame acquaintances for their crimes, sent ricin-laced mail to offices, including those of the White House, a US senator, and a judge (Associated Press, 2013; Walker, 2013). Robert Blitzer, who served as the FBI section chief for counterterrorism, had previously argued that lone wolves and extreme factions within domestic right-wing groups would be most likely to try to use CBRN weapons (Tucker, 2000: 2).

Other far-right or white supremacist groups also attempted to develop BW capabilities. There have been several arrests since 1995 of members of the Aryan Nations group who have attempted to order or steal samples of plague and anthrax (Falkenrath, Newman, and Thayer, 1998: 39). Wouter Basson, the director of Apartheid South Africa's Project Coast, which aimed to develop select agents that would target only blacks, had contact with Larry Ford, an American physician with ties to right-wing groups, who was discovered after his death to have an arsenal in his home that included the pathogens that cause cholera, typhoid fever, and brucellosis (Guillemin, 2005: 233).

Still, prior to 2001, the only recorded successful use of bioterrorism in the modern United States occurred in The Dalles, Oregon, in 1984, when members of the Rajneeshee religious sect deliberately spread salmonella in area salad bars. Their motivation was to sicken enough of the population to sufficiently diminish voter turnout in county elections to enable sect members to win office (Carus, 1997). A three-person team made the salmonella slurry with "ordinary glassware and growth media" and distributed it in restaurants using spray bottles. Although no one died as a result, 751 individuals were reported ill, with 45 requiring hospitalization (Maurer, in Maurer, 2009: 63–64, 84). The group's medical expert, a Filipina nurse nicknamed "Dr. Mengele" after the Nazi war criminal, had wanted an incapacitating agent and considered various schemes, including placing dead rodents in the area's water system and deliberately infecting people with HIV, before legally purchasing a sample of salmonella from a medical supply company in Seattle (Carus, in Tucker, 2000: 120–127). Subsequently, the group tried to assassinate critics by injecting them with lethal doses of adrenaline (Zaits, 2011).

In addition to the propensity to use biotechnology to influence political and social institutions demonstrated by the Rajneeshees, Bale and Ackerman (in Maurer, 2009: 44) argue that "cult groups with world views inspired by science fiction motifs, including the Church of Scientology and the Raelians, may warrant special attention, particularly if they promote genetic engineering or other advanced technologies." Indeed, one science-fiction doomsday cult has already done so.

Forcing the End: The Case of Aum Shinrikyo

On March 20, 1995, sarin nerve gas, a highly lethal chemical weapon, was released in the Tokyo subway system at a government center station during the 8 a.m. rush hour. Although the attack was not executed as effectively as the perpetrators had intended, eleven victims died, nearly 5,500 were incapacitated, and state resources were overwhelmed. Responsibility was soon assigned to a religious corporation, tax exempt since 1989, with the full name of "The Truth of the Creation, Sustainment, and Destruction of the Universe" (Mutsuko, 1998).

Like the Rajneeshees, cult members turned to CBRN terrorism after failures in legislative elections (Tu, 2002: 44). Aum Shinrikyo's founder, Chizuo Matsumoto, aka Shoko Asahara, had based the group's mythology on an interpretation of Buddhism that dictated that sanctity is attained through killing. Asahara claimed to have traveled forward in time to the year 2006 and spoken with survivors of a forthcoming nuclear holocaust caused by an American assault against Japan. Asahara claimed that only his disciples would be spared, but key figures in the forthcoming nuclear holocaust, including President Bill Clinton, needed to be eliminated (Religious Tolerance Council, 1998). Indeed, in addition to its multiple uses of biological and chemical weapons in Japan, Aum also had plans to attack New York and Washington, D.C. (Guillemin, 2005: 158). The cult's Ministry of Welfare was responsible for making biological weapons, as well as attempting to clone Asahara's DNA. In the early 1990s, it engaged in limited spraying of an amphetamine over crowds of people and also planned to mass-spray LSD using crop dusters (Tu, 2002: 56).

Aum was genuinely a transnational entity, with financial offices in New York (and over two hundred American followers), as well as holdings in locales as diverse as Russia, where combat weapons were purchased, and Australia, where biological and chemical weapons were manufactured and tested.[2] Aum was successful in purchasing much of this equipment legally in the United States through its corporate holdings. It also evidently received support from militant groups abroad, as evidenced by a strong anti-American and anti-Semitic content in its literature, which would seem anomalous considering the minuscule number of Jews in Japan. And while the Russian government was suspicious of Aum as a foreign proselytizing organization, the sect still managed to purchase a Russian assault helicopter through army black-market channels. It proved unsuccessful, however, in acquiring Biopreparat samples of Ebola (Brackett, 1996: 94–107).

An investigation by the *New York Times* revealed that Aum had mounted at least nine attempted biological attacks against targets in Japan, including against the Diet, the Imperial Palace, and the US base at Yokosuka (Religious Tolerance Council, 1998). Despite its transnational activities, the CIA claimed that it was unaware of Aum prior to the subway attack. Such organizations, admitted one official, "simply were not on anyone's radar screen." Despite Aum's connection to previous murders and a sarin attack on the town of Matsumoto in 1994 that killed seven, Japanese law enforcement was slow to bring charges against a religious organization (Mutsuko, 1998).

Yet Aum had attempted unsuccessfully to poison crowds at public events as early as 1990 by spreading botulinum toxin from car exhaust pipes (Falkenrath, Newman, and Thayer, 1998: 39). In 1990 and 1995, the cult attempted to disperse liquid anthrax from vans and high-rise office buildings to infect passersby. However, the slurry was too thick to be aerosolized and contained an innocuous strain of anthrax, one that ironically immunized those coming in contact with it (Preston, 2009: 233). And while Aum looked into obtaining Q fever, and Ebola specimens directly from Africa in addition to its botulin and anthrax efforts, its difficulty in mounting any successful biological attack led it to turn to chemical weapons.[3] Its failures came despite employing a research scientist with a background in genetic engineering (Maurer, in Maurer, 2009: 63–64).

Twenty-First-Century Bioterrorism

As of 2000, only three groups had actually used CBRN agents in terrorist attacks, all of which had been fairly low-tech endeavors conducted by small-scale entities (Tucker, 2000: 253). But the recognition that, prior to its far more notorious sarin attack, Aum Shinrikyo had also broken the bioterrorism taboo—albeit incompetently and harmlessly—produced a rapid revaluation of biosecurity priorities among law enforcement and national security planners internationally. The previously hypothetical scenario of psychotic or apocalyptically inclined individuals releasing deadly vectors onto an unsuspecting public, as depicted just months before the Tokyo subway attack in the 1995 science-fiction film *12 Monkeys*, had become reality. At the same time, a mounting wave of mass-casualty attacks, including various Middle Eastern suicide bombings and the destruction of the Oklahoma City Federal

Building, signaled that perhaps Brian Jenkins's 1970s dictum that "terrorists want a lot of people watching, not a lot of people dead," was becoming passé. As terrorism expert Jessica Stern (2000: 131) put it, "it is hard to deter a group that is seeking to bring on Armageddon."

One assumption that gained currency around the turn of the millennium was that after the Aum Shinrikyo cult gained attention through the use of CBRN weaponry, terrorist groups would quickly follow suit, particularly if they had religious motivations and were unconcerned with gaining widespread sympathy for their cause. "Weapons of mass destruction" would provide such actors a relatively low-cost opportunity to terrorize and traumatize a large number of people, with the public fearing contamination even if only a few actual deaths were caused by biological or chemical agents. However, this view presumes that terrorists would find these technologies more instrumentally useful than conventional attacks. Logistical difficulties aside, the perpetrators of mass-casualty attacks often find that they have alienated previously sympathetic publics, and the risk of opprobrium would presumably be greater if norms against unconventional attacks against civilians were to be violated (Bale and Ackerman, in Maurer, 2009: 11, 16, 17, 21).

And yet, the leadership of al Qaeda, which formed at the beginning of the 1990s in Afghanistan and had grown into a transnational network capable of preparing attacks on several continents by September 11, 2001, publicly stated its desire to obtain BW capabilities. In particular, the group has expressed an interest in crop dusters, the same type of delivery system used by Soviet researchers in the 1920s. When Coalition forces entered an al Qaeda facility in Kandahar, Afghanistan, they discovered a centrifuge and drying oven, along with inconclusive trace findings of anthrax and ricin (Koblentz, 2009: 200, 222, 223). After his capture in November 2001, so-called American Taliban John Walker Lindh claimed that a second wave of al Qaeda attacks would occur that December and would involve biological agents (Bergen, 2006: 349).

Abu al Masri, leader of al Qaeda in Iraq (which later became ISIS), before his death in 2010, issued a call for bioweaponeers: "We are in dire need of you [scientists]. The field of jihad can satisfy your scientific ambitions, and the large American bases [in Iraq] are good places to test your unconventional weapons, whether biological or dirty, as they call them" (Vogel, 2012: 125). In 2011, Secretary of State Hillary Clinton announced the existence of "evidence in Afghanistan that . . . al Qaeda in the Arabian Peninsula made a call to arms for 'brothers with degrees in microbiology or chemistry to develop a weapon of mass destruction'" (Garrett, 2012).

Al Qaeda official spokesman Suleiman Abu Ghaith posted an Internet article in June 2002 claiming, "It is our right to fight them with chemical and biological weapons, so as to afflict them with the fatal maladies that have afflicted the Muslims because of their chemical and biological weapons."

However, one former al Qaeda member, Paulo Jose de Almeida Santos, who joined the organization in 1990 before receiving an eight-year jail term for trying to assassinate the exiled king of Afghanistan in Rome in 1991, claimed that his suggestions to employ CBRN attacks had been declined. Santos reported that his personal offer to attempt to poison Israeli water supplies with mercury, killing individuals and destroying harvests, "was rejected." An attempt to develop a topical synthetic poison failed when the leadership sought a fatwa authorizing human testing from a sheikh who grew angered at the prospect and called them Nazis (Bergen, 2006: 337, 338, 347).[4]

Ayman al Zawahiri, the Egyptian physician turned al Qaeda leader, has gone so far as to claim that his organization only began to consider the utility of bioweapons when the United States expressed concern about the possibility (Klotz and Sylvester, 2009: 85). If this characterization is accurate, it is an eerie echo of the motivation of Shiro Ishii, the leader of Unit 731, who concluded that the Japanese Army should pursue biotech projects if the rest of the world believed that they were threatening. It could very well mean that states with bioweapon, or at least biodefense, programs are partly responsible for planting the seeds of their own security threats.

Coercion by political leaders in the Soviet and Iraqi BW programs and that of Aum Shinrikyo inhibited free exchange of data between scientists and produced pressures to present some results that justified the labor of the researchers, whether optimal or not (Ben Ouagrham-Gormley, 2014: 143). It is difficult to imagine laboratory conditions being more conducive for bioweaponeers for ISIS or other such groups.

Rather than jihadi terrorists developing their own BW production capacity, there is the probably greater prospect of government scientists stealing samples or providing know-how for financial or ideological reasons, a biotech version of Pakistani nuclear scientist Abdul Qadeer Khan. Another Pakistani scientist, microbiologist Abdul Rauf, recruited by al Qaeda, set up a weapons lab and traveled to Europe to attempt to obtain virulent strains of anthrax. In Indonesia, Jemmah Islamiyah member Yazid Sufaat, who trained as a biochemist in the United States, set up the front Green Laboratory Medicine Company to try to obtain anthrax (Koblentz, 2009: 219–221).

Between this case and the modus operandi of Biopreparat, it is likely that a number of apparently commercial biotech firms may in fact be developing weaponized biotechnology. While established scientists may not be willing to work for doomsday cults or transnational terrorist groups, the possibility exists that they may be hired by a front company operated by such an organization. Clearly the greatest threat posed by freelance experimenters is that they will impart their labors to violent non-state actors that are not subject to the deterrent threats of force that can be levied against states fixed on the map (Mutsuko, 1998).

Agroterrorism

One potential bioterrorism nexus for both private-sector and state-funded research to meet the interests of extremist or cult groups is an attack against a target's agricultural sectors. "While agriculture may not be a terrorist's first choice because it lacks the 'shock factor' of more traditional terrorist targets, an increasing number of terrorism analysts consider it a viable secondary target." Even if few people became ill or died, the economic and psychological impacts on a public unable to trust its food sources would be tremendous. Also, "the past success of keeping many diseases out of the U.S. means that many veterinarians and scientists lack direct experience with foreign diseases. This may delay recognition of symptoms in case of an outbreak, and the ability to respond to an outbreak."[5] This potential had evidently already occurred to al Qaeda researchers, whose facilities in Afghanistan "were found containing agricultural documents and manuals describing ways to make animal and plant poisons" (Congressional Research Service, 2007: 1–2).

In the vernacular of Thomas Homer-Dixon (2002), agroterrorism holds great potential as a "weapon of mass disruption." Seth Carus (1997) states that while anti-agriculture attacks by states against internal and external rivals are a distinct possibility—Saddam Hussein's Iraq admitted to developing infectants that make wheat unfit for human consumption—he argues that they would be unlikely to have a significant impact on the United States due to the size and diversity of the American agriculture industry. Still, authorities concluded that an otherwise unknown ecoterrorist entity calling itself the Breeders in threatening letters was responsible for the devastating sudden introduction of Mediterranean fruit flies into California's agriculture sector in 1989, costing between $60 billion and $120 billion in eradication and, after first causing a massive increase in wide-release aerial pesticide spraying, achieving the stated goal of ending such use after public opinion turned against it (Lockwood, 2010).

And if the object is commercial sabotage rather than large-scale famine, agroterrorism could easily have a significant impact. An outbreak of foot-and-mouth disease (FMD) decimated the British agriculture and tourist industry in 2001, and the importation of a single pig from Hong Kong infected with FMD caused the loss of $19 billion to the Taiwanese pork industry in livestock that required destruction. Additionally, the introduction of prions (transmissible protein particles that cause "mad cow disease" and related maladies and are exceedingly difficult to eliminate because they are not living organisms) into livestock would have a devastating result on industries and trade, and the peptide synthesizers that could be used to create them are "widely found throughout research and medical communities" (Preston,

2009: 308, 317–318). States, or even corporations, could therefore target each other's stocks effectively and with high degrees of anonymity.

As Segarra (2002: 4, 13) notes, many more pathogens are available that afflict livestock than humans, and a number of them, such as FMD, are not transmissible to humans, so saboteurs would not be at risk of exposure while deploying them. As of 2002, there were "no consistent minimum safety protocols or security standards for animal research laboratories in the United States," and so these types of biological agents are certainly readily available to both state and private actors. However, animal pathogens have been a human security threat since the dawn of the agricultural revolution. It is the advances in genetic engineering that offer to truly transform the nature of the threat of biological warfare, and it is unlikely that the farming industry, or terrorists armed with manure for cultivating anthrax, will be in a position to counter them.

No Cause for Terror?

Whether agroterrorism or attacks against human targets, an alarming and growing menagerie of fringe groups have expressed an interest in bioterrorism. Beyond al Qaeda and its affiliates, violent millennial groups including Aum Shinrikyo—which is still operational—and white supremacists in the "Christian Identity" movement have also been joined by radical environmentalist organizations, some of which have attacked university research laboratories processing GMOs (Gronvall et al., 2009: 434). And, as biotechnology proliferates, the ability to develop resistant, durable strains of bacteria in small facilities will become available to more groups, increasing the probability that some will be successful.

And yet, thus far, few such plans have actually been pursued. The US intelligence community announced in 2008 that fifteen groups under watch had expressed an interest in bioweapons, but only three had made any known efforts to obtain them to cause mass casualties (Koblentz, 2009: 200). Ackerman (2004: 2) notes that religion is neither a necessary nor a sufficient condition to explain a predilection to CBRN use by terrorist groups with instrumental political objectives. And apocalyptic cults will likely struggle to find highly trained technicians who can successfully implement their plans and be willing to work for such unstable employers.

For example, Ken Alibek has stated that Biopreparat researchers went to Afghanistan to interview with al Qaeda but found their prospective new employers "too creepy" (Preston, 2009: 191–195). Elsewhere, state sponsors of terrorism, such as Iran, have advanced microbiology research sectors and could potentially supply the fruits of their labors to client groups, such as Hezbollah, to act while providing the regime with plausible deniability (US House of Representatives, 2005: 7–8). This scenario assumes that state sponsors of terrorism would trust their agents enough to give them WMD and would have confidence that they would not be held ultimately responsible for their actions. With both state sponsors of terrorism and freelance researchers and technicians seemingly unlikely to trust terrorist organizations, there is some cause for optimism in this area.

Also, given the scale of major BW programs such as those of the United States and the Soviet Union, it is highly improbable that any non-state actor could marshal the resources to implement production of armaments that fundamentally threaten the security of any target country. Even if al Qaeda or an affiliated group obtained a black-market sample of a particularly virulent pathogen from a state BW program, it would still need large teams of experts and specialized equipment to be able to handle the select agents safely without exposing personnel, to store it so that it remained viable, and to disseminate it effectively (Ball et al., in Maurer, 2009: 488).

Even in the most successful case of a modern non-state actor launching attacks with CBRN weapons, Aum Shinrikyo, a well-funded transnational entity with its own technicians, was only able to make futile efforts to expose crowds to botulin and anthrax before turning instead to the chemical weapon sarin on the Tokyo subway. Aum Shinrikyo was a doomsday cult, not an organization with a state sponsor that would be reluctant to trust its agents with its stocks of (illicit) bioweapons. This is not to say that terrorists, cults, and insurgent groups will not attempt to perpetuate violence using biotechnology, but rather that it is unlikely that states will permit their BW programs to be the evident source of such attacks by third parties. Instead, as the deadliest bioterrorism incident on record indicates, the greater threat is rogue actors with training and access to select agents as a result of state biodefense programs.

THE AMERITHRAX ATTACKS

Hamid Mir, Osama bin Laden's official biographer, recalled that during an interview in November 2001, bin Laden and Zawahiri bragged about already possessing "suitcase" nuclear weapons to use as a deterrent. "[He talked about] chemical and nuclear weapons. He never mentioned the word biological because I asked this question about anthrax: 'Some people think that you are behind the anthrax attacks in America.' He laughed, and said 'We don't have any link with these attacks. Next question'" (Bergen, 2006: 348).

The leaders of al Qaeda were not the only ones caught off guard by the mailing of several envelopes of *B. anthracis* spores to media and government offices in September and October 2001. Despite growing attention to the threat of domestic bioterrorism in the wake of the Aum Shinrikyo attacks, and a number of studies identifying anthrax as the most effective pathogen widely available for such purposes, the US Centers for Disease Control and Prevention (CDC) had never considered anthrax to be a public health threat, and the FBI had only limited familiarity with the bacteria and its properties from having investigated a number of hoax attacks in the late 1990s that began after Secretary of Defense William Cohen raised the threat of an anthrax attack and announced plans to vaccinate all military personnel (Guillemin, 2005: 173). After the 9/11 attacks, President Bush began taking Cipro as a precautionary measure (Guillemin, 2011: 27).

Preparations for a bioterror attack were clearly not uniform across governmental and military agencies. While Senator Daschle would state in his memoir of the events of 2001 that "there were no human studies to rely on" in providing treatment guidelines for inhalation anthrax (Daschle and D'Orso, 2003: 168), the US Army Medical Research Institute for Infectious Diseases (USAMRIID), which provided technical assistance to health and law enforcement officials investigating the mailings, had the human experiment data provided by Japan's Unit 731 after the conclusion of World War II. And USAMRIID's senior anthrax researcher would ultimately be identified by the FBI as the sole perpetrator of the bioterrorist mailing campaign. But both the CDC and the Department of Health and Human Services (HHS) lacked access to the data on BW programs and testing. They also did not consult with USAMRIID staff who could have told them that, although the Army used a benchmark of thousands of spores for filling anthrax dispersal munitions to achieve lethal doses among half of a target population (LD50), such high concentrations were not necessary to kill individual victims.

Similarly, the FBI and Department of Defense were aware—but the CDC was not—of a Canadian study conducted the preceding year in response to hoax letters sent to Parliament which demonstrated that dispersal could also occur by leakage through the pores of a conventional envelope. Canadian officials forwarded the report when the first anthrax case was reported in Florida in October 2001, but CDC officials were too understaffed to read it until well into the investigation of the Amerithrax mailings (Guillemin, 2005: 173–174). Based on the information that it did have available, the CDC had also advised against ionization—the precaution enacted after the Capitol Hill mailings to kill anthrax by irradiating the mail—arguing that the process would only spread spores further (US House of Representatives, 2005: 37).

"We Have This Anthrax"

While Canadian bioweaponeer Frederick Banting had in the 1930s proposed anthrax mailings to strategic targets, the details surrounding the implementation of this type of biological attack remain unknown. The first anthrax infections to be reported in connection with the events of 2001 occurred among workers at the offices of the tabloid publisher American Media International (AMI) in Boca Raton, Florida. No letter of the type that delivered spores to other targets was reported or recovered from that office; however, postal facilities serving AMI were discovered to be contaminated, making it more likely that an envelope containing a significant volume of spores was delivered to AMI than the type of accidental cross-contamination that apparently killed two women in the northeast. Given that two to four other media offices were targeted, AMI would fit the established profile of high-publicity targets (US Department of Justice, hereafter USDOJ, 2010: 2).[6]

It was only when AMI employees Robert Stevens and Ernesto Blanco "happened to be admitted to the same Florida hospital for pneumonia-like symptoms and were diagnosed with inhalation anthrax that investigators had reason to suspect that an act of terrorism had occurred" (USDOJ, 2010: 11). Stevens was apparently the first to be infected and had flu-like symptoms for days before becoming seriously ill and seeking hospitalization on October 2. By the time that a spinal tap confirmed to physicians that he was suffering not from a form of pneumonia but from inhalation anthrax, which none of them had ever seen before, the infection was too far advanced to be treated (Daschle and D'Orso, 2003: 143), and Stevens died on October 5 (Guillemin, 2005: 175), perhaps the first person to perish as a direct result of modern bioterrorism.

Of the two letters to media organizations recovered by the FBI, the one addressed to the *New York Post* remained unopened until it was delivered to investigators. NBC News employee Casey Chamberlain later reported remembering receiving a threatening letter in childlike handwriting accompanied by powder around September 18 (although given that the postmark on the envelope bore that date, she presumably received it a day or two later). On September 28 she began to experience flu-like symptoms that persisted for over two weeks before being informed by supervisors on October 12 that she and an assistant to news anchor Tom Brokaw, to whom the letter was addressed, likely had cutaneous anthrax (Chamberlain, 2006).[7] On the same day, NBC News turned over the letter to the FBI, and Brokaw somberly concluded his Friday evening broadcast by recapping the situation, noting that his office staff was receiving expert medical care and doses of a previously obscure antibiotic that was becoming a household word with the unfolding coverage of the AMI incident. Holding up a prescription bottle, he intoned, "In Cipro we trust."

That same afternoon, two envelopes postmarked October 9 from the same post office in Trenton, New Jersey, that processed the recovered New York letters arrived at the US Capitol. Both letters were addressed to the correct office suites of Senators Daschle and Leahy (respectively 509 Hart Senate Office Building and 433 Russell Senate Office Building, rather than just "US Senate") and bore the same return address: the fourth-grade class of the fictitious Greenwood Elementary School in Franklin Park, New Jersey, although the zip code provided was that of the neighboring town of Monmouth Junction (USDOJ, 2010: 16). In part because a visiting delegation of indigenous Alaskans had alarmed bioterrorism-skittish office interns by attempting to give them a gift of animal fur to give to the senator, Daschle office operations wound down uncharacteristically early that Friday, one stressful month after 9/11, and the afternoon mail delivery was left unopened until the next Monday morning.

Presumably the bioterrorist hoped to generate an intense amount of publicity by infecting two prominent national political figures, or even believed that members of Congress open their own mail. As it happened, Daschle usually worked in the Senate Democratic leader's office in the Capitol rather than across the street in the Hart Building. In any case, he was never near the scene of the incident, and the task of opening the envelope fell to intern Grant Leslie, who had been quite worried about receiving an anthrax mailing despite all the reassurances given to her. Beneath the spreading cloud of powder was a piece of paper with the brief message:

9-11-01
YOU CAN NOT STOP US.
WE HAVE THIS ANTHRAX.
YOU DIE NOW.
ARE YOU AFRAID?
DEATH TO AMERICA.
DEATH TO ISRAEL.
ALLAH IS GREAT.

When four Capitol police arrived to respond to the call, none were wearing protective gear. A number of hoaxes or honest mistakes by nervous staff had cropped up since the AMI story broke, and responding to anthrax alarms had become routine procedure. By 10:30 a.m., this was already their third call of the day. The dozen staff members on the sixth-floor portion of the suite where the mail room was located were swabbed, given a single twenty-four-hour Cipro tablet as a precaution, and had their clothes collected for analysis. Otherwise the attending Capitol physicians evidently had no protocol for dealing with bioterrorism—they informed the exposed staff that they could go to the hospital if they wished, but otherwise they were free to go home. Of the twenty-five staff on the fifth-floor portion of the office, all of whom were initially told to leave without examination and only some of whom insisted on an examination that day, seven would produce nasal swabs indicating exposure. Investigators later blamed this on air circulation: it was not until after noon that the ventilation system was turned off, the system then taking an additional forty-five minutes to stop, and a number of ducts never closed. Ultimately, twenty-eight individuals in the Hart Building tested positive for inhaled spores: six emergency first responders, twenty Daschle staff, and two staffers of Senator Russ Feingold, whose office shared a wall with the Daschle suite (Daschle and D'Orso, 2003: 147–149, 151, 158, 167).

None of these congressional employees, who all received prompt and attentive medical care, were among the twenty-two victims of the attack who developed anthrax symptoms, half of whom suffered from inhalation and the other half from cutaneous infections. Among the eleven who acquired inhalation anthrax, five died: AMI employee Robert Stevens; Kathy Nguyen, a New York hospital worker who received her infection from an unknown source; Ottilie Lundgren, a ninety-four-year-old Connecticut woman infected by cross-contaminated mail; and Thomas Morris and Joseph Curseen, two US Postal Service (USPS) workers in the Brentwood facility in Washington, D.C., that processed mail deliveries to the Senate. Another ten thousand people, deemed "at risk" from possible exposure, were issued at least one day's worth of antibiotic prophylaxis. "Thirty-five postal facilities and commercial mail rooms were contaminated. The presence of *Bacillus anthracis* was detected in seven of 26 buildings tested on Capitol Hill" (USDOJ, 2010: 2–3).[8]

Overall, nine of the twenty-two victims who suffered illnesses, and seven of the eleven cases with the more severe inhalation form of anthrax, were postal workers. Four of these occurred in the Brentwood postal facility alone (Guillemin, 2005: 174–177). CDC officials had held a press conference at Brentwood on October 18, saying that they would not be there if the location was not safe (Daschle and D'Orso, 2003: 168). However, as reported by the NGO Judicial Watch in December 2002, the manager of the Brentwood mail processing plant had written in his diary the previous day that the facility was contaminated (Guillemin, 2005: 178). The dearth of information shared by management with employees at Brentwood came to be emblematic of poor coordination in response to the attacks and of the fact that risk and resources were not evenly distributed across government employees.

On the day that the letter was opened, some of the Daschle staff had not trusted the assurances of the Office of the Attending Physician that they could not have been exposed and did not require testing, and one staff member circulated a petition demanding nasal swabs that the staff signed during the hours of quarantine (Kane, 2006). While all staff would probably have been tested within a couple of days once the scope of the contamination had become apparent, quite possibly the early detection prevented some of the two dozen who tested positive for exposure from contracting inhalation anthrax, but not in their nasal passages. It is difficult to imagine that a similar protest would have been effective under the conditions described at Brentwood.

In the case of the Senate staff, CDC recommended that a sixty-day course of antibiotics and no vaccinations would be sufficient to kill 99 percent of spores. However, as nasal swabs had indicated that some Daschle staff had three thousand times more spores in their respiratory system than what was considered to be a lethal dose, various members of the group argued that following the CDC protocol could still leave them with more than enough surviving spores to prove fatal. The Senate majority leader, sensitive to the well-being of his staff and their concerns, used his clout to sideline CDC and place officials from the Bethesda Naval Hospital in charge of staff care after they recommended a more aggressive treatment protocol based on their existing guidelines for treating exposed Marines (Daschle and D'Orso, 2003: 168). Thus, a group of largely Caucasian political appointees was far more successful than the mostly African-American working-class USPS laborers in obtaining effective advocacy among the competing organizations of the federal bureaucracy claiming jurisdiction over the case.[9]

The congressional employees exposed in the attack had the benefit of an immediate realization that they had been exposed. But they also benefitted from the patronage of the most powerful member of the US Senate and could be assured that they were receiving attentive care. The reopening of their renovated suite in the Hart Building the following spring—following a remediation treatment which Annie Jacobsen notes in *The Pentagon's Brain* was suggested by DARPA (Temple-Ralston, 2015)—was marked by a party in which they were presented with souvenir T-shirts, humorous news-clip compilation videos, scrapbooks, and certificates.

By contrast, affected postal workers formed the Brentwood Exposed support and publicity group

> to deal with the deaths, illness, stress, and reactions to antibiotics caused by the anthrax letters, for which no federal agency took responsibility. That no perpetrator had been found contributed to the community's loss of trust in law enforcement and government. Their workplace took two years to decontaminate, and many workers were reluctant to return.[10] (Guillemin, 2005: 178)

"Sociopathic with Clear Intentions"

The investigation of the attacks, concluded by the FBI over seven years after they had occurred, was at least as significant in providing information about the state of homeland security measures and forensic microbiology as it was in ultimately naming a suspect. On the day after the Daschle letter was opened, an interagency briefing for congressional leaders characterized the spores in the Hart Building as weapons grade, which pointed to a skilled scientist and a state BW program (Guillemin, 2005: 175).

Expert technique was evident in concentrating the spores from slurry. Forensic evidence demonstrated that the *B. anthracis* powder sent in the Senate letters was more refined than the specimens recovered from the earlier mailings to the media offices, indicating that the perpetrator had been perfecting the process over the intervening three weeks (Koblentz, 2009: 206).

However, the initial characterization of an attack using a weaponized pathogen, repeated by some officials responsible for the investigation and cleanup
and by political and media figures, was incorrect:

> The spore particles had a mass median diameter between 22 and 38 microns.
> They exhibited an electrostatic charge, showed no signs of genetic engineer-
> ing, and were non-hemolytic, gamma-phage susceptible, antibiotic and vac-
> cine sensitive, and devoid of aerosolizing enhancers (e.g., fumed silica, ben-
> tonite, or other inert material). These characteristics were and are inconsistent
> with weapons-grade *anthracis*. (USDOJ, 2010: 14)

Further, despite some public speculation about the possible role of Saddam Hussein's Iraq, handwriting experts quickly concluded that the author of
the letters naturally wrote in Latin rather than Arabic script, and American
military scientists soon determined that the *B. anthracis* spores were of domestic origin: "This strain, known as 'Ames,' was isolated in Texas in 1981,
and then shipped to USAMRIID, where it was maintained thereafter. Another natural outbreak of Ames has never again been recorded" (USDOJ,
2010: 3).[11]

That "only 15 U.S. and three foreign laboratories were known to possess
the Ames strain of anthrax prior to the attacks" all but foreclosed the possibility of a foreign terrorist attack.[12] That "the perpetrator almost certainly
came into contact with aerosolized anthrax spores in committing the crime,
and was probably protected against an anthrax infection by vaccination and/
or antibiotics" was further indication of a source within the American biodefense establishment—who else would have access to a rare and virulent
pathogen and appropriate prophylactic treatments? Finally, additional investigations by swabbing postal infrastructure for spore concentrations

> allowed investigators to identify a heavily contaminated blue street-side box
> located across the street from the main entrance to Princeton University. . . .
> After several months of investigation, investigators concluded in August 2002
> that this was the box from which all of the attack letters were mailed. (USDOJ,
> 2010: 12, 15, 16)

At this point, as the investigation focused on the search for a probable
suspect, attention turned to Fort Detrick, Maryland, where USAMRIID
worked, ostensibly on purely defensive research, with Ames strain *B. anthracis*. A public leak revealed that the FBI had determined USAMRIID researcher Steven Hatfill to be a "person of interest" in the investigation after
"numerous persons" contacted the FBI to suggest him as a suspect. In addition to having filled multiple prescriptions for Cipro in 2001,

> while working as a researcher at USAMRIID from 1997 to 1999, Dr. Hatfill
> had virtually unrestricted access to the Ames strain of anthrax, the same strain

used in the 2001 mailings. Dr. Hatfill also appeared to know the intricacies of conducting successful anthrax dissemination by mail, although it was not uncommon for those in the bio-defense community to develop such scenarios for training exercises. (USDOJ, 2010: 6)

However, Hatfill was ultimately excluded as a suspect (and subsequently won a multi-million-dollar personal damages lawsuit against the federal government) when,

in 2007, after several years of scientific developments and advanced genetic testing coordinated by the FBI Laboratory, the Task Force determined that the spores in the letters were derived from a single spore-batch of Ames strain anthrax called "RMR-1029." . . . Identifying, classifying, and testing for the [identifying] genetic mutation had never before been accomplished with this bacterium. In fact, the tests needed to conduct such an analysis of *Bacillus anthracis* did not even exist in 2001. . . .

RMR-1029 had been created and maintained by Dr. Bruce E. Ivins at USAMRIID. This was a groundbreaking development in the investigation. It allowed the investigators to reduce drastically the number of possible suspects, because only a very limited number of individuals had ever had access to this specific spore preparation that was housed at USAMRIID. . . . Early in the investigation, it was assumed that isolates of the Ames strain were accessible to any individual at USAMRIID with access to the biocontainment labs. Later in the investigation, when scientific breakthroughs led investigators to conclude that RMR-1029 was the parent material to the anthrax powder used in the mailings, it was determined that Dr. Hatfill could not have been the mailer because he never had access to the particular bio-containment suites at USAMRIID that held the RMR-1029. . . .

Investigators learned that Dr. Ivins was alone late at night and on the weekend in the lab where RMR-1029 was stored in the days immediately preceding the dates on which the anthrax could have been mailed. Before the anthrax mailings, Dr. Ivins had never exhibited that pattern of working alone in the lab extensively during non-business hours, and he never did so after the anthrax attacks. When confronted, he was unable to give a legitimate explanation for keeping these unusual and, in the context of the investigation, suspicious hours. (USDOJ, 2010: 5–7, 24)

As the Department of Justice was preparing to indict Ivins for bioterrorism and murder in 2008, he committed suicide by taking a massive overdose of Tylenol. The FBI was satisfied that Ivins was the sole perpetrator and closed the case in 2010. His colleagues at USAMRIID, however, claimed that the evidence against him was circumstantial and that his suicide was motivated by facing the same ruinous costs and public scrutiny suffered by the previously named Hatfill (Klotz and Sylvester, 2009: 111). One scientist declared it was "absolutely not" possible that Ivins could have been culpable, equating blaming him with tying a gun shop owner to a murder weapon. He further claimed that Ivins would not have had access to protective equipment to avoid spreading spores around the facility (Shane, 2010).

However, Ivins, who began his work with *Bacillus anthracis* at USAM-RIID in 1980 and had more than fifty publications on the subject, had enlisted the help of a colleague in late 2001 to disinfect his work area from what he described as an accident involving spore transfers—an incident he never reported to the FBI. But more pertinent to the investigation was motive:

> According to his e-mails and statements to friends, in the months leading up to the anthrax attacks in the fall of 2001, Dr. Ivins was under intense personal and professional pressure. The anthrax vaccine program to which he had devoted his entire career of more than 20 years was failing. The anthrax vaccines were receiving criticism in several scientific circles, because of both potency problems and allegations that the anthrax vaccine contributed to Gulf War Syndrome. Short of some major breakthrough or intervention, he feared that the vaccine research program was going to be discontinued. Following the anthrax attacks, however, his program was suddenly rejuvenated. (USDOJ, 2010: 8, 26)

Even this evidence is circumstantial and does not provide evidence of culpability. However, there was one additional factor concerning Ivins that led the FBI to conclude that he was the perpetrator, and one that carries tremendous implications for homeland security efforts and biosecurity planning more generally:

> Between 2000 and his death in 2008, Dr. Ivins had been on varying doses of antipsychotic and anti-depressant medication. It is clear from his e-mails dating back to 1998 (the earliest e-mails investigators were able to obtain from USAMRIID computers) that Dr. Ivins was suffering from significant mental health problems at the time of the anthrax attacks. In e-mails sent in 2000 to Former Colleague #1 and Former Colleague #2, two women on whom he was admittedly fixated and reliant, he expressed concerns about "delusional" thoughts he was having and feared that he was becoming increasingly mentally disturbed. . . .
>
> In the month before his suicide, his homicidal tendencies became more pronounced, as he posted violent messages on the Internet regarding a reality TV star and made death threats during a group therapy session. One of the mental health providers who was present when Dr. Ivins made these threats noted in publicly filed court papers that Dr. Ivins had "a history dating to his graduate days of homicidal threats, actions, [and] plans," and that a prior psychiatrist "called him homicidal [and] sociopathic with clear intentions." . . .
>
> He had a penchant for going on long drives to mail letters and packages from distant post offices, often using a pseudonym when doing so, thereby disguising his identity as the mailer. [13] (USDOJ, 2010: 9, 10, 42)

With one counselor diagnosing Ivins with paranoid personality disorder, the most significant question concerning the investigation would not seem to be who was responsible or why they did it, but why such an individual was permitted to remain in a position of authority within the US Army's biodefense complex and allowed unsupervised access to select agents. As David Willman (2011) observed, "one need not be convinced of Ivins's guilt to realize he should not have been allowed anywhere near the Army's anthrax, let alone given unrestricted, 24/7 access to live spores for nearly 28 years." Yet, as Willman learned through a Freedom of Information Act request, despite his well-known psychiatric problems, "Dr. Ivins was never evaluated by USAMRIID for mental fitness."

Perhaps more importantly, why did none of his colleagues, who had witnessed an array of disturbed and threatening behavior, notify the FBI at the outset of the investigation? Ivins provided recommendations and testing expertise to investigators for years while the case was underway, apparently tampering with some results along the way (USDOJ, 2010: 44–45). But he only became a suspect when new means of genetic testing made his involvement with the spores used highly probable; prior to that there was evidently a code of silence maintained about his behavior within USAMRIID. Obviously no such omertà was practiced among the "numerous" colleagues who suggested Hatfill as a suspect. Extrapolating this behavior to the entire biotech industry yields troubling conclusions.

SECURITIZING BIOTECHNOLOGY

After the 9/11 and Amerithrax attacks of 2001, the United States dramatically increased spending in the name of what had previously been a vaguely hypothetical term of interest to terrorism specialists: homeland security. Some of the initiatives had begun in the late 1990s in response to the Tokyo subway and Oklahoma City Federal Building attacks of 1995. But it was the first serious manifestation of bioterrorism in America that provided the impetus to reprioritize programs of the type that Bruce Ivins worked on that had apparently lost their urgency with the conclusion of the Cold War. This development, when combined with the unfolding biotech revolution simultaneously occurring in the commercial and academic sectors, meant a large increase in the number of laboratories processing select agents, and an even greater number of individuals with access to them, in some cases with haphazard safety precautions or no effective oversight. The biotech sector has undergone a form of securitization since the Amerithrax attacks, but it is less clear whether the expansion of the biodefense industry in the name of homeland security is answering the potential threat of bioterrorism or raising the probability of future attacks.

Biodefense Programs

Biosecurity efforts in the United States date back to Washington's decision to inoculate the Continental Army against smallpox in 1777 and continued through the end of the nineteenth century with the efforts of the Army Medical School to eradicate endemic diseases such as typhoid and yellow fever. Walter Reed determined that the latter, like the malaria that afflicted much of the United States at the time, was mosquito borne and that "the Army had the tools to compel civilians to change in ways an elected government in the United States could not. Such requirements included: covering and screening water cisterns, implementing anti-mosquito patrols, oiling puddles, and digging ditches to drain marshes" (Marble, 2010).

Other efforts continued through World War II, as evidenced by instructional films produced by Walt Disney Studios, such as *Defense against Invasion* (1943), which truly securitized health by depicting harmful germs as enemy saboteurs and white blood cells as well-armed paratroopers. Still, Leitenberg and Zilinskas (2012: 2) note that Cold War civil defense education programs did not raise the threat of biowarfare to the public as they did with the possibility of nuclear attack.

Still, prior to 2001, education campaigns were generally the most that the publics in states that had developed BW programs could expect. Instead, it was assumed that mutual deterrence would prevent biological warfare from becoming a reality. It was not until bioterrorism by actors perceived to be immune to deterrence (because they were apocalyptic or willing to commit suicide and because they had no fixed location against which to retaliate) that large-scale biodefense became a security priority in many states.[14]

President Bill Clinton, consulted by Craig Venter (at the time affiliated with the United States' Human Genome Project), was particularly concerned about the prospect of genetically modified select agents being used to thwart biodefense measures and in 1997 directed Secretary of the Navy Richard Danzig to begin developing countermeasures against potential lines of mass-casualty BW terrorism.[15] Secretary of Defense Cohen subsequently announced that all American military personnel would be required to be vaccinated against anthrax, a controversial initiative that led to some soldiers accepting a dishonorable discharge rather than be subjected to what they argued was an untested treatment. While only a very small percentage of recipients experienced serious side effects, severe illness and death did result in some cases, and the perception that the Pentagon was blocking information about the harmful effects of the vaccine led some to charge that the inoculations were responsible for the "Gulf War syndrome" afflicting some veterans of the 1990–1991 campaign against Iraq, although others also attributed it to a claimed or theorized exposure to Iraqi chemical or biological weapons (Guillemin, 2005: 161–165).

Perhaps because the countermeasures required for biological attacks were outside of the traditional security threats against which Defense was accustomed to preparing, after World War II, biodefense in the United States became a public health responsibility of the Department of Health and Human Services (Smith, 2011: 663–696). After the Aum Shinrikyo ricin attack, President Clinton determined that the Federal Emergency Management Agency would coordinate responses to CBRN attacks, which further cast biodefense as a domestic policy area (Smith, 2014: 111).

"Integrated training exercises began in selected major cities in 1998 that involved the coordination of military, federal and local authority responses to simulated bioterrorist attacks," and "for the 1999 budget, President Clinton was successful in arranging $300 million in addition to the annual $1 billion already approved to fight bioterrorism"(Pogrebin, 1998). This included the creation of the National Laboratory Response Network, from which stockpile CDC obtained enough Cipro and Doxycycline to treat thirty-three thousand patients for potential anthrax infection in 2001, even though it was not known how many of them had actually been exposed (Guillemin, 2005: 174).

Within the first five years after the attacks of 2001, the United States had spent more than $20 billion on biodefense measures (US House of Representatives, 2005: 2). The Bioterrorism Act of 2002, the first comprehensive congressional response to this particular challenge to public health security, mandated increased surveillance of drinking water, water supplies, and food sources and required further stockpiling of drugs and supplies and accelerated research on pathogens. The Bush administration, in the context of its buildup toward war in Iraq and its suspicion about the survival of Iraq's illicit BW program, also insisted on pushing forward in late 2002 with mandatory smallpox vaccinations for half a million military and government employees who might be deployed to the Middle East, followed in 2003 by optional vaccinations for millions of "frontline" domestic civilian emergency first responders (Guillemin, 2005: 180–183).

By 2004, the United States had allocated nearly $30 billion to homeland security, with the largest outlay for the new BioShield program (Guillemin, 2005: 183). "The [Project BioShield Act of 2004] set aside $5.6 billion over 10 years for [HHS] to develop and acquire medical countermeasures—diagnostics, medicines and vaccines—to prepare for threats to public health from unconventional events" (Zigmond, 2008).

But the program immediately ran into difficulty in finding suppliers, in part because its goal was to procure drugs that did not normally otherwise have a commercial market (Klotz and Sylvester, 2009: 165). Critics quickly complained that the program produced little of value and lacked direction because it was being influenced by competing biotech firms lobbying to include their products as essential (Bardach, in Maurer, 2009: 226–227), with others noting that stockpiled vaccines would only be useful against particular select agents, and only if they had not been modified to counteract those particular preventatives or treatments (US House of Representatives, 2005: 40).

By 2006, the only bid for a contract to produce a "next-generation" anthrax vaccine had yielded underwhelming results and the contract was terminated (although a Danish manufacturer subsequently received a contract for smallpox vaccine production), and billions of dollars that had been appropriated for BioShield remained unspent (*Defense Daily*, 2008). In response, the program updated late that year to focus on promoting the creation of new broad-spectrum antivirals and antibiotics (Klotz and Sylvester, 2009: 166), although in 2001, the Transformational Medical Technologies Initiative intended to develop broad-spectrum vaccines ended with no evident success (Smith, 2011: 67).

At the same time, Congress responded to the shortcomings of BioShield by passing the Pandemic and All-Hazards Preparedness Act, thus creating the Biomedical Advanced Research and Development Authority (BARDA), which was granted the latitude to determine threats and invest in appropriate countermeasures and R&D. It also created an assistant secretary for preparedness and response at HHS to coordinate the planning efforts of various governmental agencies and coordinate logistics in an emergency (Bardach, in Maurer, 2009: 226–227; Zigmond, 2008).

Monitoring Systems

In 2001, prior to the major terrorist attacks of that year, the Committee on Opportunities in Biotechnology for Future Army Applications noted that battlefield detection of CBRN material would soon become faster and more reliable with the development of small, inexpensive

> biological sensors, or *biosensors*, [that] may be defined as devices that probe the environment for specific molecules or entities through chemical, biochemical, or biological assays. The targets can be airborne, in liquids, or in solid materials. Biosensors may involve any or all of the following functions: detection, capture, concentration, derivitization, and analysis of samples. . . . Ranging from several square centimeters to the size of a computer chip, a small sensor that can perform all of the functions normally carried out at the laboratory bench is sometimes referred to as a "laboratory on a chip." (Committee, 2001: 16)

The new prioritization of homeland security led to a rapid adoption of biosensors for domestic applications, including the BASIS mobile aerosol collectors used by the Department of Homeland Security (DHS) and the USPS Biohazard Detection System (Labov and Slezak, in Maurer, 2009: 157). But what both of these programs shared with the more ambitious Bio-

Watch network (deployed nationwide in 2003 by DHS to detect minute quantities of pathogens used in bioterrorist attacks) is that they were "labor intensive, costly, and the resultant data is significantly time-delayed," as samples needed to be periodically collected manually and then sent to labs for analysis, a process often taking longer than twenty-four hours (Biesecker, June 30, 2009).

Other limitations plagued BioWatch as well. By not providing data in real time, it could not support interdiction efforts against actors disseminating select agents. It was also limited by the genetic codes contained in search databases, so it would not have detected novel or engineered pathogens (Gronvall et al., 2009: 440). Additionally, by merely detecting the presence of genetic material, the sensors could not determine if samples represented live or dead pathogens, of if they were occurring in such low levels that exposure would not present a health risk. The initial program, consisting of Gen-1 and then Gen-2 sensors, was most useful in providing the ability to determine whether an attack had already occurred, information still important for locating contaminated individuals before they begin to manifest symptoms (Labov and Slezak, in Maurer, 2009: 142–143).

In 2010, DHS began deployment of fully automated Gen-3 sensors that would cost less per unit in annual maintenance and that would also "continuously monitor the air around the clock for certain biological agents, do automated analysis throughout the day and provide automated alerts via a communications network." The devices also scan for all pathogens on threat lists (Biesecker, June 30, 2009), rather than simply collecting samples and having lab technicians scan for priority threats beginning with anthrax.

> Under Gen-3, DHS plans to have the machines not only collect the air samples but also perform the analysis automatically and provide an alert within four to six hours if a potential threat has been encountered. Moreover, given the relatively quick collection and analysis times, the Gen-3 machines will also be designed to work indoors in places such as airports and subway systems. (Biesecker, 2008)

In the first decade of the program, the United States spent "about $1 billion to deploy BioWatch detection equipment in more than 30 U.S. cities and at major spectator events, including Super Bowls and national political conventions." However, during this time, health officials recorded at least fifty-six false threat alerts, including one that nearly necessitated the cancellation of Barack Obama's presidential nomination acceptance speech in 2008. DHS has noted that these were not false positive readings because they did detect some organism, even if it did not turn out to be a threat. Nonetheless, the high level of alerts led the CDC to decide not to issue medications based solely on a BioWatch reading, the Federal Aviation Administration canceled plans to install sensors in airports, and the City of New York had existing sensors removed from the subway system as a result (Willman, 2012).

Little advancement had occurred by 2015, with the dozen models of biosensors available to DHS still incapable of full-spectrum scans. Although the DHS Science Directorate partnered with private companies that had received $350 million in federal contracts, the dozen types of sensors that had been developed "can be clunky and difficult to operate for anybody but a highly skilled lab technician" and require one to three days to deliver results. Some companies that produced reportedly superior devices found their contracts canceled without explanation, suggesting that political factors affect biodefense as much as they do all other forms of government contracting (Markon, 2015).

Contemporaneous with the initiation of BioWatch in 2003, the CDC launched its BioSense program to attain real-time surveillance data. BioSense collects data from a wide array of hospitals, including Veterans Administration facilities and private institutions to look for unusual patterns that might indicate disease outbreaks, including particular symptoms that might indicate the use of certain bioweapons, in a process known as syndromic surveillance (Davis and Ringel, in Maurer, 2009: 311). Information gathered by health agencies at various levels of government also includes trends in 911 calls, school absences, and emergency room reports. Without effective sensors, physician diagnoses would be the first indicators of an intentional biological release (Labov and Slezak, in Maurer, 2009: 156, 165).

As CDC described it, "the BioSense Program goal is to support a national surveillance network through which healthcare organizations, public health, Health Information Exchanges (HIEs), and other national health data sources are able to contribute to the picture of the health of the nation" (CDC, *BioSense Overview*, 2009). To this end, it established the National BioIntelligence Center in 2004 to coordinate monitoring in real time (Bradley et al., in CDC, 2005: 15). While various local governmental agencies began syndromic surveillance prior to 2001, the magnitude of patient data collection involved in BioSense is a recent development with a direct bearing on how other individual medical privacy and patient rights (addressed in greater detail in the next chapter) might come into tension with collective rights to homeland security.

Homeland Security and Biotechnology

Between these new programs and the expansion of existing BW research programs, in the decade after 9/11 and Amerithrax the United States spent nearly double the cost of the Manhattan Project in current dollars on biodefense (Klotz and Sylvester, 2009: 98). In the latter category was the National Interagency Biodefense Campus (NIBC), created in 2002 to centralize R&D

in biotechnology across health, military, and homeland security agencies. Among its components are the National Biodefense and Countermeasures Center as the principal research institution (Bardach, in Maurer, 2009: 227), the Biological Threat Characterization Center intended to develop defenses against new types of bioweapons, a potentially dual-purpose activity (Klotz and Sylvester, 2009: 90–91); the National Institute of Allergies and Infectious Diseases; and USAMRIID (NIBC, 2009).

The inclusion of this last institute in the NIBC in 2002 is somewhat curious considering that it was at this time that it became clear that someone on the inside was involved in the unfolding Amerithrax case. Additionally, the campus is located at Fort Detrick, Maryland, the very facility from which the modified Ames strain spores originated. While the sparse NIBC website provides a brief list of accomplishments consisting primarily of children's vaccine developments, it is clear that the scope of its activities extends beyond biodefense and epidemiology.

The edifices of the biomedical wing of American homeland security are built—literally—on the foundations of its industrial BW program begun for World War II. Thus programs begun for entirely offensive purposes have been rededicated to ostensibly purely defensive research and mobilization. But, for better and for worse, they also contribute to the securitization of biotechnology as an expanding array of biotech concerns are classified under the umbrella of national preparedness. The Obama administration, continuing to build on the Bush administration's inaugural DHS and related programs, proposed a 6 percent increase in homeland security spending during its first year in office, providing DHS alone with a discretionary budget of $42.7 billion (Biesecker, February 27, 2009).

However, despite the direction of billions of dollars and large portions of the federal bureaucracy toward "homeland security," many potential targets in the United States remain highly vulnerable. There are several means by which terrorists could launch a BW assault against the American homeland. These include aerosol dissemination, anti-agriculture efforts, and targeting water and food supplies. Alibek, who stated that he expects *B. anthracis* spores to be used in future attacks because of their proven durability, argued that bioterrorism differs from conventional forms of terrorism that rely on bombings because it can continue for extended periods of time, as with Amerithrax. Costs associated with bioterrorism are tremendous as well, given that approximately six grams of spores caused billions of dollars in remediation costs. Additionally, he notes that the fixed defenses on which so much money have been spent, such as BioShield, are futile because any parties launching an intended mass-casualty biological attack will ensure that their select agents are resistant to the antibiotics stockpiled for them (US House of Representatives, 2005: 6, 28, 37, 46).

At the least, bioweapons will continue to present effective weapons of mass distraction for terrorists. Drinking water supplies are likely safe from serious harm because of the dilution effect and routine chlorination. Milk and a few other food supplies are also pasteurized and subject to recalls and media coverage that would prevent most people from ingesting contaminated products. But the mere attempt in either case would be effective in wreaking fear and confusion and leading to consumer boycotts of suspect goods. Aum Shinrikyo's attacks killed a dozen people and the Amerithrax incidents produced five fatalities, but the mass panic and cleanup costs associated with each incident demonstrated the efficiency of small amounts of exotic weaponry and how they can be used effectively against high-use infrastructure. Therefore, particularly considering that "technology barriers to bioweapons seem to be eroding far faster than for other forms of WMD," Maurer (in Maurer, 2009: 96–97, 103) argues that "for now the most worrisome WMD threat probably involves bioweapons and toxins."

RESEARCH SECTOR SECURITY AND PROLIFERATION

In 2011, four men in their sixties and seventies with white supremacist militia ties were arrested for plotting to obtain castor beans in their plot to "rescue" the state of Georgia from the Obama administration. They had been recorded planning to spread ten pounds of ricin by aerial dissemination from the open windows of a car traveling on the interstate, in addition to conventional attacks against government facilities using guns and explosives. One of the plotters was an employee of the USDA, another a CDC contractor with a top secret clearance (Reilly, 2011).

If bioterrorism remains as a significant homeland security threat, it is important to discern the sources from which it would originate in order to try to prevent future attacks. To date, bioterrorist plots worldwide have been attempted almost exclusively by domestic actors, many with access to government BW programs and their networks of research labs.

The research sector's secrecy and bioethics are therefore of vital import to international security. Outside of the governmental facilities, attempts at self-regulation through the establishment of industry-wide professional standards and ethics began with the 1975 Asilomar Conference, but few external controls are in place. This is particularly the case for the physical security of specimens, "where dangerous material is often kept in unlocked refrigerators, and where background checks on workers are rare" (Cohen, 2002).

Preston (2009: xxii, 189, 195) estimates that worldwide there are fourteen thousand people in four hundred different government and contracted labs, ostensibly working on defense-related projects, who probably have access to "select agents" to use as weapons-grade pathogens. And "unfortunately there are hundreds of germ banks and laboratories world-wide that lack adequate security." A single spore of anthrax could be used to grow an unlimited amount, and dozens of companies manufacture fermenters, freeze-driers, and centrifuges that individuals can use in processing purloined samples. In addition, a growing number of individuals have access to labs processing such select agents or potential bioweapons, as well as the capacity to develop them acting alone.

Lab Security

In the United States, the CDC has established four levels of biosafety for laboratories, ranging from BSL1, which involves organisms that do not cause disease in humans, to BSL4, the designation for facilities that process pathogens for which no treatment is readily available, such as the Marburg and smallpox viruses (Federation of American Scientists, 2010). Even after 9/11, some BSL4 facilities were unguarded to the degree that individuals without clearances could walk right up to, and in some cases into, the buildings during security audits (Commission, 2008: 3). Beyond the risk of stolen samples was the threat of vehicular bomb attacks demolishing buildings and releasing pathogens into the atmosphere. And even without external tampering, conditions are potentially precarious at many biotech facilities.

At the beginning of the 2010s, approximately 232,000 employees worked in biotechnology labs in the United States (Pollack and Wilson, 2010), including estimates by the Biotechnology Industry Association and the FBI of 1,500 biotech firms and 22,000 labs holding samples from outbreaks and epizootics. In many cases, security at these facilities, particularly those of small private firms, is loose from an administrative as well as a physical security standpoint, with inconsistent enforcement of regulations and little preventing front companies, such as those operated by al Qaeda affiliates described earlier in this chapter, from purchasing them and absorbing their specimens. Some of these facilities are operated by academic institutions whose research could subsequently be used to refine existing BW programs and arsenals (Maurer, in Maurer, 2009: 77–80). And developments in synthetic biology may permit researchers, perhaps inadvertently, to create entirely new organisms against which no defenses are available (Pollack and Wilson, 2010).

The National Institutes of Health (NIH) established regulatory safeguards against this eventuality in 1974, soon after the advent of recombinant technology, creating a Recombinant DNA Advisory Committee that informs the policies of the NIH Office of Biotechnology Activities. NIH has claimed not only that "compliance with the NIH Guidelines is mandatory for investigators at institutions receiving NIH funds for research in recombinant DNA," but also that "they have become a universal standard for safe scientific practice in this area of research and are followed voluntarily by many companies and other institutions not otherwise subject to their requirements" (NIH, 2010).

Klotz and Sylvester, however, argue that young faculty members engaged in cutting-edge research often have difficulty securing NIH funding, so they instead pursue grants from the burgeoning defense sector, often meaning involvement with more hazardous specimens. Since 9/11, a great deal of money has been available for research on biodefense—which often involves the very select agents against which defenses are being constructed—and universities and private enterprises have been encouraged to build BSL facilities. Klotz and Sylvester (2009: 4, 5, 9, 94, 177) argued that BSL4 programs signal hostile purposes to foreign observers because pathogens like the Ebola and Marburg viruses are unlikely public health issues in the United States.

Ebola did become an international security concern during the West African epidemic of 2014. However, the four cases reported in the United States consisted of two volunteers who had spent time in affected areas in Africa and two hospital workers who had attended one of them without adequate precautions. That no fellow airline passengers or other members of the public were ever infected would indicate that even the most virulent nonendemic diseases do not ordinarily pose a domestic security threat. Interestingly, at the 2015 DARPA Biotechnology Program Office launch, Defense-connected researchers mentioned that they plan to use their Living Foundries biobrick manufacture program to develop counters to future Ebola outbreaks (Rothman, 2015).

The net effect of defense funding of biosecurity is therefore the securitization of biotechnology. Additionally, universities and other research centers have historically been sites of bioweapons R&D, both in the United States and the Soviet Union. Other seemingly commercial labs, such as pesticide manufacturers, have also been used to cover BW programs (Alibek, 1999: 42, 52). The distribution of advanced BW research to academia and private entities is concerning given available information about the state of security in such institutions. For example, during 2003–2004, HHS tested universities for compliance where research with select agents was being conducted. It determined that eleven of fifteen sites suffered operational weaknesses that could have resulted in theft or accidental loss of dangerous material (Klotz and Sylvester, 2009: 135). As these are the types of institutions that the secretary of defense has suggested for DARPA partnership expansion, the potential for the proliferation of military-purposed biotechnologies is also likely to increase.

At the same time, the number of BSL4 labs in the United States tripled from five to fifteen in the decade after 9/11, with "some 400 research facilities in the United States . . . authorized to store and handle select agents, and nearly 15,000 individuals have been approved to work with them. [In response,] the U.S. government has sought to foster the development of a 'culture of security awareness' within the life sciences community to prevent the misuse of biology for warfare or terrorism" (Commission, 2008: 25). Another proposal is that, rather than relying on simply screening individuals when they are hired, attentive managers could make efforts to develop "collegial work relationships that support recognition of significant changes in colleagues. A secure laboratory workforce is one in which crises that may lead to potentially dangerous changes in personnel are noticed and addressed" (Gronvall et al., 2009: 435).

In 2002, the CDC launched its Select Agent Program that required researchers to report work with any of forty-two select agents, updating 1996 antiterrorism legislation that only mandated reporting if lethal pathogens were transferred between facilities. Researchers "also had to submit the names of employees with a legitimate need for access to the pathogens to the secretary of HHS and the office of the US Attorney General" to check for individuals who might be restricted for security purposes. "These regulations applied to open nongovernmental laboratories, with the presumption that the government was monitoring its own classified research on dangerous pathogens" (Guillemin, 2005: 184).

"However, scientists in academia and industry generally view the Select Agent Program as an unnecessary burden rather than as an important means of preventing bioterrorism" (Commission, 2008: 25–26). Additionally, many facilities, particularly small private biotech firms that are not dependent on federal funding, do not fully observe these regulations and may be employment magnets for individuals who have difficulty obtaining security clearances (Klotz and Sylvester, 2009: 100–101). The effect could conceivably be driving those who are perhaps the likeliest perpetrators of future biological attacks out of the oversight of regulators.

Even at the source of the Amerithrax incident, a military research facility, security problems persisted. In 2009, several months after charges emerged that senior researcher Bruce Ivins was creating Amerithrax with uncataloged samples, a USAMRIID inventory uncovered 9,220 vials not in the database, including specimens of Ebola and botulinum neurotoxins (Smithson, 2011: 243). After temporarily suspending research to address the accounting system, the institute's commanding officer then acknowledged "that the probability that there are additional vials of BSAT [biological select agents and toxins] not captured in our . . . database is high." Additional audits discovered that security measures to prevent theft of materials were lacking in other facilities, with "very few of the 400 institutions" that possess bioweapons having comprehensive video monitoring of work areas, regulations that at least two persons be present when select agents are handled, or psychological screening or monitoring of personnel (Sigger, 2009).[16]

In 2008, the CDC conducted an assessment of the BSL4 facilities in the United States and discovered that twelve of the fifteen labs had in place no more than three out of fifteen safety controls evaluated. Two labs lacked either internal or external surveillance cameras or armed guards posted at all public entrances. "One lab even had a window that looked directly into the room where BSL4 agents were handled. In addition to creating the perception of vulnerability, the lack of key security controls at these labs means that security officials have fewer opportunities to stop an intruder or attacker. [The CDC Division of Select Agents and Toxins had] approved the security plans by the two labs lacking most key security controls" (US Government Accountability Office, September 2008: i, 4).

Nor is the phenomenon, and potential source of biological threats, unique to the United States. An audit of ninety-four laboratories at twenty-two different research facilities in Denmark found similar results. Although "sensitive agents" were present in nearly half of the locations, "in 81 percent of pathogen-containing facilities, pathogens were not routinely and centrally accounted for." During the investigation, "it was demonstrated that the backgrounds and identities of insiders were rarely checked and that they could have gained access to both pathogen inventory lists and freezers in many facilities." Nearly a quarter of BSL3 facilities (containing lethal but treatable pathogens like anthrax) had no locks on their storage freezers, and another made keys to such locks accessible to its cleaning crew. Fourteen of sixteen public institutions also permitted cleaning crews with no security clearances unaccompanied access to pathogen storage areas. Only half required personal identification equipment to access the laboratories, with two generally leaving doors unlocked during the day and another one leaving them unlocked at night as well (Bork et al., 2007: 62–68).

Beyond the possibility of malfeasance is the certainty of accident. Data is somewhat limited in this area, but with numerous instances of mishaps involving select agents reported in a variety of institutions, it is highly likely that a number of other incidents have gone unreported as well. Indeed, some have come to light only through external investigations.

> A survey done by the Bureau of Labor Statistics in 2006 found that the rate of workplace injury and illness in corporate scientific research laboratories was well below the average for all industries. [However, the] survey included labs in industries like information technology as well as biotechnology, and excluded labs handling the most dangerous pathogens. . . . One study, reviewing incidents discussed in scientific journals from 1979 to 2004, counted 1,448 symptom-causing infections in bio labs, resulting in 36 deaths. About half the infections were in diagnostic laboratories, where patient blood or tissue samples are analyzed, and half in research laboratories. . . . The casualties include an Agriculture Department scientist who spent a month in a coma after being infected by the E. coli bacteria her colleagues were experimenting with. An-

other scientist, working in a New Zealand lab while on leave from an American biotechnology company, lost both legs and an arm after being infected by meningococcal bacteria, the subject of her vaccine research. [In 2009,] a University of Chicago scientist died after apparently being infected by the focus of his research: the bacterium that causes plague. (Pollack and Wilson, 2010)

Two years later, another researcher from the same lab was hospitalized with a skin infection from *Bacillus cereus* (Kaiser, 2011).

At a lab at Texas A&M University, it went unreported that an employee was exposed to brucella, a BW agent, with the incident only coming to light when a colleague was sickened by another specimen and misdiagnosed. Three other lab workers at the same university failed to report that they had been infected with Q fever, and researchers in Seattle also contravened the law when they did not report infection by tuberculosis. A New Jersey facility reported losing three plague-infected mice in 2005 before announcing that it had been an accounting error, but the same facility then lost two infected mice corpses in 2009. In Britain, the exposed corpses of dead test animals and pathogens that apparently escaped in other instances have transmitted infection to livestock, and Australian researchers working on sterilizing mice inadvertently made mousepox lethal enough to kill previously vaccinated animals (see chapter 2). A likely cause of these occurrences is not only that the proliferation of biosecurity facilities since 2001 has created more opportunities for errors, but that the multitude of new employees also lack the judgment that comes with experience: 97 percent of new hires made since 2001 had no reported prior experience handling pathogens (Klotz and Sylvester, 2009: 12, 114–115, 122, 127–129).

In 2015, Defense officials announced that a US Army lab had inadvertently sent live anthrax samples to over fifty laboratories in the United States and three foreign countries, giving every individual who had access to the samples at every recipient institution the opportunity to grow their own stocks from the spores (McLeary, 2015).

According to Alibek, the Soviets experienced similar problems under the Biopreparat regime: a researcher, and then his coroner, died after being pricked by syringes of Marburg virus–infected blood, another worker nearly died from a cutaneous anthrax infection, and Alibek himself was exposed to tularemia while cleaning a lab spill, an incident that he did not report and survived only by treating himself at home with illegally obtained pharmaceuticals. The decades-old grave of a colleague on Rebirth Island indicated that such occurrences were nothing new, and civilians were periodically exposed as well: fishermen in the Aral Sea were exposed to plague when the wind unexpectedly shifted (a similar incident occurred during American nuclear tests in the Marshall Islands in the 1950s), and atypical instances of plague were reported in that vicinity of Central Asia for some time. The deaths of dozens of residents of Sverdlovsk occurred because a faulty air filter on a BW plant was not fixed properly, and the strain of *B. anthracis* responsible had originally been cultivated from rats infected by a leak at another facility more than twenty years earlier (Alibek, 1999: 16, 63–78, 84, 105, 124–131).

The Dual-Use Dilemma

As with a number of labs in the United States today that process hazardous materials, many of these Biopreparat facilities, some purporting to be manufacturing plants, operated in dense urban centers. And, as noted, many of them used the same equipment and specimens that pharmaceutical or agribusiness would employ. It is not only this Janus-faced nature of biotech that has posed difficulties in evaluating its impact on national defense and international security, but the fact that many of these developments have emerged from the private sector with ambiguous applications. Whether biotechnologies are employed for WMD or commercial products—or both—depends on the actor wielding them. For example, "botulinum toxin, ten thousand times more lethal than the nerve gas VX, is marketed under the name Botox." Indeed, many ingredients in bioweapons, the research necessary to understand them, and the equipment used to prepare them, also have legitimate medical and therapeutic uses as well. Restricting the availability of some biotechnology might therefore hamper vital medical research, and policy makers must confront the trade-offs (Koblentz, 2009: 5, 64).

"Drug makers, responding to competition from cheap generic medications, are moving beyond the traditional business of making pills in chemical factories to focus instead on vaccines and biologic drugs that are made in vats of living cells" (Pollack and Wilson, 2010). Fletcher and Allen (2007: 5) note that "firms that are developing breakthrough biotechnologies are focused on patent and property rights as well as the intrinsic value of the scientific discovery. Such actors generally resist constraints imposed by societal institutions."

In part this is due to the financial pressures on pharmaceutical firms: typically only one out of tens of thousands of drug candidates will advance to the commercial market, and by the time it clears testing, the seventeen-year patent for a drug is usually halfway to expiration. Likewise, these high costs mean that large publicly held pharma companies focus on developing profit-maximizing "blockbuster" drugs, leaving the small independent labs to conduct biodefense research. Additionally, corporate espionage is on the rise, resulting in an estimated $3 billion in lost sales during the 1990s and demonstrating the lack of adequate safeguards on biotech research (Koblentz, 2009: 72, 231).

Additionally, "bacteria are cultivated identically whether they are intended for industrial application, weaponization, or vaccination" (Alibek, 1999: 53). And the fact that the equipment used in BW programs is virtually indistinguishable from commercial processes has not changed in the twenty-first century. In the survey of Danish bioresearch facilities, "dual-use biological equipment was found in two of three (66%) pharmaceutical companies, in one of five (20%) public research institutions, [and] in one of seven (14%) universities" (Bork et al., 2007: 62–68). The production of vaccines, which rely on specimens that have been killed, first requires the production of large quantities of those pathogens (Koblentz, 2009: 68).

For these reasons, there is wide availability of both the technology and technical expertise to create BW programs. Worldwide, there are perhaps more than one thousand research groups and more than ten thousand people with the training to resynthesize a virus and release it. Tens of thousands of DNA synthesizers sit in laboratories, equipment that could be purchased for about $1 million in start-up costs. Finally, production cost barriers are low: one study showed that fourteen million lethal doses of anthrax could be created in a twenty-square-foot lab with about $220,000 in equipment and using open-source information, including out-of-print microbiology textbooks. And the means are available worldwide: over two hundred labs in sub-Saharan Africa hold anthrax and plague samples from humans, and many of these facilities are in countries hostile to the West and outside of the sight of its investigators (US House of Representatives, 2005: 16, 17, 38).

Experts on India and China report that their respective governments and private biotech sectors lack the regulatory or surveillance capacity to ensure biosecurity in their burgeoning number of dual-use and hacker labs (Garrett, 2012). However, even less-developed Asian states, including Bangladesh, Nepal, and Bhutan, now have their own BSL3 labs (Smithson, 2011: 245). And dozens of companies worldwide manufacture fermenters, freeze-driers, and centrifuges used in biological research (Preston, 2009: 189).

In response to these factors, but initially created in the 1980s to monitor dual-use equipment for chemical weapon manufacture when those armaments were being used during the Iran-Iraq War, the Australia Group formed to harmonize export controls on sensitive technology. Critics charge that it is a club of forty-one developed nations intervening to retard the progress of developing countries in promoting their own biotech sectors for disease prevention and self-sustaining agriculture (Guillemin, 2005: 194; Koblentz, 2009: 234). Other dual-use export regulations also cover synthetic DNA, which has been used in recent experiments to make polio and the pandemic Spanish flu (Maurer, in Maurer, 2009: 77). The synthetic development and laboratory dissemination of such pathogens is presently completely legal in each country in which the research has been conducted (Garrett, 2012).

In 2004, the secretary of health authorized the creation of the National Science Advisory Board on Biosecurity (NSABB), "a cross-government committee to address the dual-use conundrum, finding a way to deter terrorist or other malicious use of scientific discoveries without impeding the pace of basic discovery and invention." While the US government delegated national security decision-making authority to academics, NSABB members initially established that their role was only to recommend self-censorship in particular cases when sensitive information was being reviewed for publication. It took no action when an article presenting the sequencing of the Black Plague was published (Garrett, 2012).

In 2012, in a decision that argued that the risks stemming from publication outweighed any potential benefits, the NSABB recommended censoring a publication that would have detailed how researchers had successfully manipulated the H5N1virus to enable airborne transmission between mammals. In so doing, it compared this research in bird flu manipulation, involving a pathogen that produced 60 percent fatality rates in humans who had been able to catch it only from birds, to early research on atomic weapons and acknowledged the potential for an "unimaginable catastrophe" (Vastag and Brown, 2012).

The technologies are already available globally, and advances over the past forty years mean that technical barriers to once exotic procedures have fallen away. "What in the past would have required a massive, state-run biological weapons program with substantial infrastructure and technical support can today be accomplished by state and non-state actors alike in buildings the size of a standard garage using easily acquired, off-the-shelf equipment and technically proficient personnel" (Preston, 2009: 9). Even if Aum Shinrikyo's team failed in its efforts because it lacked specific knowledge, "one competent microbiologist" with a graduate education and a few thousand dollars could establish a functional BW program (Preston, 2009: 188, 234). "Today, thousands of biologists worldwide possess the requisite skills [to create bioweapons] and more are trained every day (most often at US universities). Finally, they do not require rare infrastructure; some BW can be produced by small terrorist groups almost as easily as through national biological warfare programs" (Block, in Drell, Sofaer, and Wilson, 1999: 42).

Garage Hackers

Maurer (in Maurer, 2009: 86) notes that commercially available micronizers can produce particles of one to ten microns, and in their advertising material, their manufacturers "boast that they can be operated by 'anyone . . . in their garage.'" Non-state actors with interests in these technologies have been quite busy utilizing such machinery in the past decade, with individuals referred to as "garage hackers" operating autonomously with small pieces of equipment (US House of Representatives, 2005: 30). In urging the implementation of bioethics plans, the International Committee of the Red Cross (2002) warned against the "creation of viruses from synthetic materials, as occurred this year using a recipe from the Internet and gene sequences from a mail order supplier."

In her *Wall Street Journal* article "In Attics and Closets, 'Biohackers' Discover Their Inner Frankenstein," Jeanne Whalen (2009) recounts how

> private individuals, presumably merely hobbyists, purchase DNA samples online, and then attempt to genetically modify E. Coli and other pathogens in home refrigerators or garages using equipment purchased on eBay for tens of dollars. . . . Synthetic DNA, made from material normally found in the nucleus of cells, is readily available for commercial purchase. Some are simply mixing test tubes to develop luminescent bacteria. Other individuals have spent a reported $20,000 on projects to develop alternative biofuels.

Of course other possibilities exist as well. A functional polio virus genome has been assembled by research teams using DNA sequence information, and the sequence for smallpox is available on the Internet (Klotz and Sylvester, 2009: 22). Whereas it was possible to restrict the development of traditional bioweapons by limiting access to specimens, it is no longer possible to prevent individuals with no oversight from engineering pathogens when all that is needed to create them is genetic sequences rather than actual samples. While this is clearly a frightening prospect, some observers have suggested attempting to take advantage of the situation by treating it as analogous to the threats and promises of open-source computing. By promoting openness, it may be possible to bring more people, perhaps aspiring entrepreneurs, into the biodefense sector, with potentially far more individuals out there working for the common good than harm, and ultimately legions of citizens designing vaccines (*The Economist*, "And Man Made Life," May 22, 2010).

BIOSECURITY IN THE TWENTY-FIRST CENTURY

Proposals for "wiki-biodefense" come nearly full circle from the original state bioweapon and biodefense programs initiated nearly a century earlier, in which military planners recognized that BW would be most effective against civilian populations and infrastructure and that the most efficient counters would be public health education campaigns. In the intervening time, virtually all BW victims have been civilians and, except for the war crimes of Unit 731 in Manchuria, in peacetime. Unforewarned individuals have been infected with anthrax by accident from Russia to Japan to the

United States, and it seems probable that future victims of bioterrorism and advanced biowarfare will also be civilians rather than combatants.

Terrorist groups have until now been unlikely perpetrators of bioterrorism. Because most pathogens have short shelf lives, such groups would need to build large stockpiles for mass attacks while also avoiding accidents or somehow hiding them. And given the failure of Aum Shinrikyo, which had trained experts and a large budget, it seems unlikely that strategic terror organizations would risk significant resource outlays on producing material hazardous to their members that would have little likelihood of succeeding (Maurer, in Maurer, 2009: 77, 85).

Still, advances in DNA sequencing and the wide availability of genetic information mean more resources available for any groups attempting to create their own BW programs.

Because of periodic episodes of attempted bioterrorism, states will not dismantle their biodefense R&D programs in the foreseeable future. And yet it has been material leaked, intentionally in the Amerithrax incident, from state biotech programs that has killed more individuals than all non-state actors have ever done using CBRN weapons. And the expansion of these programs and their interface with private labs increases the odds of mishaps and misdeeds.

While it is comforting to believe that even destitute scientists would not work for crazy cults or violent terrorist groups, precedent does not support this hope, as the case of Aum Shinrikyo demonstrates. At the same time, the growing number of government labs for biodefense are ironically probably the chief source of biotechnology threats. Personnel screening, even when employees publicly exhibit violent and irrational behavior, has obviously not prevented some individuals from misusing select agents. And even rigorous screenings would not prevent competent lone-actor garage hackers from carrying out attacks, even just against personal acquaintances, if they chose to do so.

Although it has evolved since the twentieth century, the threat to the human security of individuals posed by state biotech programs will continue to exist for decades. But most citizens would probably not want them to disappear because of the continuing threat of bioterrorism. Without a perfect solution or the ability to put the biotech genie back in the bottle, citizens of states with biodefense programs will have to weigh the consequences of different eventualities, at which point biosecurity becomes a question of ethics.

NOTES

1. During Daschle staff briefings by health authorities, it became clear that NIH officials were surprised by the vulnerability of the highly processed strain of *B. anthracis* used in the

attack to even penicillin as a treatment. This would seem to lend credence to the FBI theory that the perpetrator's main objective was to call attention to the potential threat of bioterror attacks and obtain research funding rather than to cause fatalities.

2. Guillemin (2005: 159) reported that Aum had thirty thousand members in the former Soviet Union and another twenty thousand in Australia.

3. Tucker (2000: 7) reported that Aum members developed Q fever infections.

4. Without human test subjects, the project was discontinued when animal experimentation proved ineffective. Al Qaeda researchers had hoped to obtain a pig, which has a similar metabolism to a human, in an effort to continue, but were unable to locate any in the Muslim lands of South Asia (Bergen, 2006: 338).

5. The Amerithrax case provides evidence that the symptoms produced by pathogens that health officials are unaccustomed to treating may well initially be misdiagnosed as a more common malady.

6. Letters sent to the *New York Post* and NBC News were recovered with Trenton, New Jersey, postmarks dated September 18, 2001, one week after 9/11, although all recovered letters were dated September 11. Employees or relatives at ABC News and CBS News in New York were also reported to have been infected at around the same time as the NBC employees who contracted cutaneous anthrax. It is presumed that letters were also mailed to these offices and AMI at the same time.

7. Most of the New York media employees and their relatives who were infected had gone to a doctor, but three with cutaneous anthrax were misdiagnosed with having a spider bite, and another was told her lesion was a reaction to her acne medication (Guillemin, 2011: 49–58).

8. Ironically, spores from the Leahy letter—which had been misdirected to the State Department mail room—cross-contaminated a diplomatic pouch bound for Yekaterinburg, Russia, formerly known as Sverdlovsk, the site of the accidental *B. anthracis* release from a Biopreparat facility that killed dozens in 1979 (Guillemin, 2005: 178).

9. For a lengthier list of interagency disputes involving the CDC, other executive branch agencies, and the Senate majority leader, including how Daschle personally intervened with the Secretary of Defense and the president to obtain vaccinations for his staff, see Daschle and D'Orso, 2003. For a description of the ailments affecting survivors of the attack exposed while working for the US Postal Service, see Klotz and Sylvester, 2009: 65.

10. Some of my former colleagues have indicated to me, fairly, that the relatively small and cohesive nature of our staff contributed to more efficient monitoring of physical and emotional health among the Daschle staff than other offices were able to enjoy. Also, I do not wish to trivialize the fact that they (and I) were also victims of bioterrorism, and many suffered emotional health impacts and the losses of both personal effects and faith in authorities—the ultimate target of all terrorist attacks. See North et al. (2005) for additional data.

11. A clerical error led the specimen to be identified as originating in Iowa rather than being found in a dead cow in Texas. As with all prior BW work, researchers were interested in the hardiest and most virulent select agents available. Given its specimens, the availability of equipment, and enough trained personnel to preclude the immediate identification of a suspect, if USAMRIID was not violating the BWC by engaging in intentionally offensive research, it at least provides evidence of its indistinguishability from purely defensive studies.

12. Korbitz (2010) notes that, prior to 2001, it was common for microbiologists to share cultures, so individuals or entities outside of USAMRIID may have had ample opportunity to generate Ames spores.

13. Ivins also reportedly had a fixation with coded messages and had written letters to colleagues using the DNA codons ACGT to signify hidden meanings. The FBI speculated that coding the name of a colleague with whom he was fixated, or possibly an anti–New York message, was why some *A*s and *T*s were boldfaced in the letters sent to New York media offices and why penicillin was misspelled "penacillin" (USDOJ, 2010: 58–60).

14. An exception was Israel, where all citizens have had gas masks since the Scud missile attacks launched by Iraq in 1991 (Maurer, in Maurer, 2009: 90).

15. Smith (2015: 63) argues that Danzig consciously extolled the necessity of biodefense funding primarily as a means to maintain department budgets that faced reductions in the post–Cold War period.

16. Fort Detrick, the launch point of Amerithrax, has elected not to implement the two-person rule (Willman, 2011).

Chapter Five

Just Warfare and Bioethics

The preceding chapters have covered a wide variety of historical and plausible future effects of biotechnology on both international balances of power and on individual human security. Although concerns over biological warfare, and later bioterrorism, stemmed directly from great power bacteriological weapons programs and then the possibility that the material could be obtained or duplicated by violent non-state groups, there are many more facets to the roles of biotechnology in international security. These include the spread of epidemics by troop movements, the use of genetic engineering and biomimetics to augment warfighters, and the levels of risk introduced to civilian populations by both the bioweapon and biodefense industries created for the stated purpose of defending them.

All of these aspects of biotechnology raise ethical questions that can neither be easily resolved nor dismissed. BW programs have already produced tremendous costs in lives and financial resources. They have also delivered on the promise of saving lives in wartime, and they offer the double-edged promise of providing more humane and reversible weapons. But all of the potential promises and perils of biotechnology can be examined in the context of ethical decisions that must be confronted not only by scientists, but by the political and military leaders who elect to fund and implement biotechnologies—or who refrain from doing so.

Bioethics, as defined by the *American Journal of Bioethics*, is the "study of moral issues in the fields of medical treatment and research. The term is also sometimes used more generally to describe ethical issues in the life sciences and the distribution of scarce medical resources." The field has been influenced by the ancient Greek Hippocratic oath for physicians to "do no harm," by the violations against human subjects by totalitarian regimes, and by subsequent advances in reproductive and genetic research (Caplan and McGee, 2004).

Such a discipline might at first glance appear to be philosophical compared to the operational concerns of military planning and homeland security. But warriors, statesmen, and administrators have long wrestled with how to appropriately integrate biotechnology with military force. More broadly, they have also been forced to address whether certain types of unconventional weapons and tactics are permissible, particularly their use against noncombatants, an avenue that has already been pursued by at least one leading state military. Therefore, with scientific developments offering an ever-expanding range of roles for biotechnology that outstrips the ability of doctrine to keep pace with every new advance, scholars and practitioners of national security can at least turn to established principles on the ethics of warfare to provide precedent.

BIOLOGICAL WEAPONS AND THE ETHICS OF WARFARE

Although international laws, or at least reciprocal agreements, limiting BW do not appear until after World War I, history is replete with both admonitions against the use of toxins in battle and the establishment of norms that otherwise limited their use. International norms against BW, although they have not universally held, bear special examination in attempting to explain why they have not been used more widely. In *The Chemical Weapons Taboo*, Richard Price examines another category of twentieth-century unconventional arms with a long lineage of historical antecedents. He finds that, like BW, CW were typically not used even when opponents lacked the capacity to respond in kind, nor in situations in which military planners expected them to have great utility (such as eliminating Japanese defenders hiding in caves during World War II). As with many new technologies of warfare (including crossbows, submarines, and machine guns), CW were initially condemned as immoral by status quo powers because they provided a potential asymmetric advantage to their adversaries. It is this type of challenge to the power-based hierarchy in the international system rather than any particular destructive capacity or cruelty in their effects that makes CBRN weapons normatively unpalatable (Price, 1997: 2–6). As Hedley Bull (2002: 48) noted, nuclear weapons introduced into the international system the possibility of the final fulfillment of ultimate Hobbesian anarchy, in which "the weakest has strength enough to kill the strongest."

Historical Norms Concerning Biological Attacks

The recognition of this quality of BW has no doubt colored the estimation of their value throughout history. As described in chapter 1, biotechnologies have been implements of warfare used by various cultures since the dawn of civilization. Military and political theorists of the ancient world and medieval period proposed proscribing bioweapons on normative grounds. But they also recognized, much as in the international society of the nineteenth and twentieth centuries, that the strategic appeal of unconventional weaponry militated in favor of developing guidelines for appropriate offensive and defensive strategies:

> The Brahmanic Laws of Manu, a code of Hindu principles first articulated in the fifth century B.C., forbade the use of arrows tipped with fire or poison. Written in India a century later, Kautilya's *Arthashastra*, one of the world's earliest treatises on war and realpolitik, advocates surprise night raids and offers recipes for plague-generating toxins, but it also urges princes to exercise restraint and win the hearts and minds of their foes. (Tharoor, 2009)

Based on a number of proclamations by key strategists, it appears that the Roman Empire developed a strong norm against biowarfare. "The Roman military historian Florus denounced a commander for sabotaging an enemy's water supply, saying the act 'violated the laws of heaven and the practice of our forefathers'" (Tharoor, 2009). There are few recorded attempts of efforts made to poison other armies, and the jurist Valerius Maximus stated firmly that "war is waged with arms, not poison" (Guillemin, 2005: 3). However, this statement was made in reaction to the poisoning of wells by Germanic tribes desperate to slow the advancing legions (CBWInfo, 2005), providing evidence in support of the hypothesis that CBRN are most strongly condemned when they are used to foil hegemons.

The condemnation of enemy tactics as violating martial tradition is hardly surprising, but it is less clear why the Romans appeared to largely reject biowarfare. The legions were known to poison water supplies with botulin from decaying animals during sieges (Roberts, 2003: 15), so the prohibition was not complete. While the explanation may lie in Rome's particular military or political culture, it is also likely that its overwhelming advantage in conventional force usually made it unnecessary to resort to less certain unconventional weapons. Also, the role of the professional poisoner Locusta as a state assassin (as described in chapter 1) made it easy to gender poisoning as the work of deceitful women and devoid of the martial glory conferred by physical strength (Price, 1997: 23).

Tu Mu, a ninth-century commentator on Sun Tzu's *The Art of War*, argued that it was imperative to camp upriver from the enemy to deny them the opportunity to poison your water supply (Sun, 2009: 166). Niccolò Machiavelli (1521: 67–68), in his own version of *The Art of War*, noted that strategies for victory included luring the enemy to defeat by glutting them on food and wine, with some generals from the classical period poisoning them for good measure. However, military victory through poisoning and guile appears to have been generally proscribed during the medieval period. "German gunners in the late Middle Ages pledged not to use 'poisoned globes' or any poison since to use such devices was considered unjust and 'unworthy of a real soldier.' . . . In 1675 French and German armies [also] agreed not to use poisoned weapons against each other" (Roberts, 2003: 25).

However, there is evidence that these norms were not taken to apply to the Other, which is to say groups outside of what was defined as the civilization or international society of the time. For instance, witnesses to the use of poison (burning rags dipped in a liquid that produced a noxious smoke) in 1456 by defenders of Belgrade against Turkish invaders commented that such a practice should never be used against fellow Christians, but that it was nonetheless effective and may be permissible against other Muslim forces in the future (Price, 1997: 36).

Given the clash of civilizations frame imparted by some on counterterrorism efforts against transitional Islamists and the deployment of Western troops as occupation forces throughout the Islamic world, it is of no small consequence to consider whether policy makers or individual troops are more willing to employ biotechnology against human targets viewed as alien. [1]

The widespread norm of poisons being inappropriate weapons against rival soldiers also often has not appeared to extend to civilians. The Roman willingness to poison the water supply of population centers while generally avoiding attempts to poison opposing armies was echoed in the twentieth century by the Hitler regime, which studiously avoided the use of chemical weapons against Allied forces but experienced no qualms about their use against civilians in concentration camps. Japan in the same period used BW against noncombatants, both civilians and prisoners of war, but made no recorded successful attempts to use them as offensive weapons on the battlefield.

Still, the ban on BW use against conventional enemies gathered strength as modernity made biotechnology and the germ theory of pathogenicity more accessible. The poisoning of water supplies by desperate, retreating Confederates in the American Civil War "led to the issuance by the US Army of General Order No. 100 which stated 'The use of poison in any manner, be it to poison wells, or food, or arms, is wholly excluded from modern warfare'"[2] (Roberts, 2003: 83).

Subsequently, the norm rapidly cascaded internationally. No participants stated their opposition during the 1874 Brussels Conference that banned poisoned projectiles. And although the United States was the sole opponent of the ban on CW-loaded artillery shells enacted by the Hague Conference of 1898, it supported prohibitions on poison or poisoned weapons by the 1907 Hague Conference, and it championed the 1925 Geneva Protocol that banned both CW and BW (Price 1997: 8, 19; Roberts, 2003: 84).

Chemical and biological weapons were therefore proscribed before they were ever used on the modern battlefield, a development that coincided with the advent of modern norms for the acceptable conduct of warfare. And because civilians were not targets of CW attacks in World War I, Price (1997: 12) argues that the global public never became accustomed to them or inured to them. The norms against chemical and germ warfare were so entrenched by World War II that President Roosevelt could declare in 1943 that "I have been loath to believe that any nation, even our present enemies, would or would be willing to loose upon mankind such terrible inhumane weapons. . . . Use of such weapons has been ruled out by the general opinion of civilized mankind" (Krickus, in Walkin, 1986: 414).

Of course, the United States had by this point produced tons of select agents to be used in disease bombs, as had the other major belligerents in the war. The research scientists who served as architects of the American BW program opposed first use primarily on strategic rather than moral grounds, arguing that it would initiate a destructive tit-for-tat exchange: "The likelihood that bacterial warfare will be used against us will surely be increased if an enemy suspects that we are unprepared to meet it and to return blow for blow." Similarly, Admiral William Leahy reported that he warned Roosevelt in 1944 that "the reaction can be foretold—if we use it, the enemy will use it." And he also added that the use of BW "would violate every Christian ethic I have ever heard of and all of the known laws of war. It would be an attack on the noncombatant population of the enemy" (Guillemin, 2005: 29, 60).

In 1947, during the early Cold War period in which it continued to build its industrial-scale BW program, the United States submitted a draft resolution to the United Nations adding biological arms to the category of WMD, which already contained nuclear and chemical weapons (Guillemin, 2005: 12). The following year, the testimony during the Nuremburg trials of Holocaust perpetrators who had committed deadly and dehumanizing medical experiments led to the 1948 UN Genocide Convention that prohibited acts targeting particular groups for physical or emotional harm, as well as the prevention of human reproduction, which would preclude bioweapons that attacked particular ethnicities. The destruction of noncombatants in death camps, as well as by firebombing and nuclear attacks in World War II, had by this point rendered WMD morally unacceptable because they inevitably killed innocent civilians (Christopher 1994: 210).

The US Army *Land Warfare Manual*, written in 1956 but updated and used through the present day, requires soldiers to respect "principles of humanity and chivalry," with the latter quality historically viewed as inconsistent with BW. It further expressly recognizes that the Geneva Protocol restricted the use of poison, poisoned weapons, and herbicides. While it maintains the right to abrogate the protocol if other states do so, it notes that the reservation of the United States does not, however, reserve the right to retaliate with bacteriological methods of warfare against a state if that state or any of its allies fails to respect the prohibitions of the Protocol. The prohibition concerning bacteriological methods of warfare which the United States has accepted under the Protocol, therefore, proscribes not only the initial but also any retaliatory use of bacteriological methods of warfare. In this connection, the United States considers bacteriological methods of warfare to include not only biological weapons but also toxins, which, although not living organisms and therefore susceptible of being characterized as chemical agents, are generally produced from biological agents. All toxins, however, regardless of the manner of production, are regarded by the United States as bacteriological methods of warfare within the meaning of the proscription of the Geneva Protocol of 1925.

And yet, shortly thereafter, it announces that "the United States is not a party to any treaty, now in force, that prohibits or restricts the use in warfare of toxic or nontoxic gases, of smoke or incendiary materials, or of bacteriological warfare" (US Department of the Army, 1956: i–xi). Given this ambiguity, and a century's worth of data indicating that major powers continue to develop BW capabilities while condemning their use, it seems likely that doctrine alone will not prevent the evolution of biotechnology for the purposes of warfare throughout the twenty-first century.

Bioweapons and the Just War Tradition

And yet the military doctrines of the United States and other Western nations are based on centuries of tradition and developed philosophy and, even if the precepts have not been fully upheld, they still shape and constrain the logic of appropriateness for BW usage. Most prominent among them is the just war theory advanced in the early seventeenth century by the Dutch philosopher and political official Hugo Grotius. In the Grotian conception that laid a foundation for international law, advanced in *On the Law of War and Peace* (1625), war is an occasionally necessary evil that is bound by constraints on its legitimate purpose and conduct. With the goal of limiting the evils of warfare, just war theory requires viewing all individuals as possessing equal universal human rights regardless of their particular religion or national origin (Krickus, in Walkin, 1986: 412).

Among the tenets of the just war doctrine of *jus in bello*, or the legitimate conduct of warfare, is *mala in se*, which prohibits combatants from using weapons or methods that are "evil in themselves."[3] Paul Christopher includes in this category "weapons whose effects cannot be controlled, like biological agents," presumably meaning that deliberately released bacteria and viruses are likely to cause collateral damage. He notes elsewhere that noncombatants are typically regarded as "innocents" and that attacking them is therefore a war crime (Christopher, 1994: 169, 196).[4]

It is equally unacceptable that uninvolved civilians may be harmed, even well after the termination of the conflict, by stray BW vectors. Whereas the effects of CW would typically be confined to the battlefield before they could disperse more widely, biological agents might be dispersed over wider areas, and some agents could survive for years (Krickus, in Walkin, 1986: 416). Reports of high rates of plague in Japanese-occupied China well after the termination of the activities of Unit 731, and similar reports near former Biopreparat facilities in Central Asia, lend credence to this view. Such logic has also successfully fueled the International Campaign to Ban Landmines.

But what of BW that do not injure bystanders but only their intended targets? New direct-effect weapons offer that potential. And, as both Price and Christopher note, there is no obvious reason why CBRN weapons should be viewed as more morally abhorrent than guns or conventional explosives that also kill and maim. In the just war tradition, however, there is a ban on inflicting "unnecessary" suffering, which is measured by the degree to which it continues after the belligerent ceases to be a combatant. This would include poisoned weapons that produce infections that debilitate or kill the target after they have been wounded and ceased to be a threat (Christopher, 1994: 106–107, 201, 205).[5]

The humane protection of wounded soldiers presumes that they have effectively ceased to be combatants and pre-dates twenty-first-century battlefield medicine and survival rates, let alone soldier enhancement. It does not consider that belligerents might have the capacity to treat themselves so effectively that they may either immediately resume their missions or return to the field in a repaired and even augmented state. It also provides no doctrine on whether their opponents have the right to strategically prevent them from doing so in the name of saving their own people from continuing attack. Army personnel who have expressed a preference for new medical biomaterials that would permit wounded soldiers "to complete a mission or, at a minimum, reduce the impact of the injury on the unit's mission" (Committee, 2004: 11–14) must consider whether such abilities would prompt adversaries to ensure that they use sufficiently lethal force to prevent the need to face rejuvenated troops.

Lin, Mehlman, and Abney (2013: 41) note that the number of aerial drones and ground robots employed by the US military increased by forty and one hundred times respectively during the first decade of the century, with little consideration given to how the new emblems of US power are perceived abroad. They contend that international audiences view them as cowardly ways of fighting and therefore dishonorable, increasing antipathy and therefore risk to the United States and its military personnel even while they make the United States more likely to go to war because of lower upfront human costs. They further argue that augmented soldiers are liable to similarly be viewed as dishonorable, both inhuman and inhumane, creating both a negative strategic value and a justifiable argument for attacking them with excessive force.

Just war tradition does permit the type of reciprocal attacks that have been the cornerstone of state biowarfare doctrines since their initiation in the interwar period. Reprisals are acceptable against violators of CBRN weapons norms, provided that the attack was a deliberate policy and not an unapproved act by a renegade individual (a General Ripper in *Dr. Strangelove*), because they might be an effective means of preventing additional violations (Christopher, 1994: 196). Such action would be permissible as a double effect, in which an otherwise objectively evil act was not intended as an end in itself, but rather as a means to reducing levels of destruction. Reprisal attacks can therefore constitute *jus ad bellum*, or legitimate reasons for war, and some scholars argue that even the killing of noncombatants and the use of poison weapons are acceptable in this context (Walzer, 1977: 153, 215).[6]

The view that retaliatory attacks using unconventional armaments are justifiable was evident during World War II, with Churchill warning Hitler over open radio broadcasts in 1942 that the use of chemical weapons against the Soviet Union would be treated as equivalent to their use against Great Britain and justify even larger chemical air raids against military targets in Germany in response (Guillemin, 2005: 46, 54). A similar logic of coercion underlay the NATO containment of the Soviet Union during the Cold War. And across the length of that conflict, "military decision makers argue[d] that unless they [were] prepared to wage wars of all kinds—conventional, nuclear, guerrilla, chemical/biological—there [was] no hope of deterring enemy attacks on the free world." Additionally, the rising importance of counterinsurgency operations also led military planners to reconsider the potential use of chemical and biological weapons that had been mooted in the Pacific theater in World War II, including "tear, vomiting, or nerve gas to flush out rebels while reducing their own casualties in turn" (Krickus, in Walkin, 1986: 416–417).

Nonlethal Weaponry

Still, norms against CBRN weapon usage held during the Cold War, with a possible significant exception being the mass deployment of defoliants by the United States in Vietnam and the disputed reports of Soviet mycotoxin use in Afghanistan. Although apparently a marginal view, proponents of unconventional weapons continued to argue that they would be more humane tools of warfare than conventional arms and would save lives in counterinsurgency operations:

> Non-lethal weapons are a class of weapons unique to CB [chemical/biological] warfare. Nonlethal CB weapons pose a new type of problem for proponents of

the just war doctrine. . . . We must ask whether specifically nonlethal CB agents are more humane than conventional and nuclear armaments. Shall we wage a completely non-lethal war because of these agents? . . . Suppose the United Nations had access to non-lethal weapons and were able to use such weapons in the Congo conflict, for instance, the Katangese army and the white mercenaries could have been subdued while the number of casualties would have been reduced by one-quarter to one-half. The question can be put this way: if such weapons are available to us, are we immoral in sticking to conventional warfare? (Krickus, in Walkin, 1986: 420–421)

However, both the Geneva Protocol and the BWC prohibit the use of either lethal or nonlethal biological weapons, and the 1977 Additional Protocol I to the Geneva Protocol prohibits states from targeting civilians with nonlethal weapons or intentionally causing superfluous injuries such as blinding. It also prohibits weapons that result in a widespread, long-lasting, or severe effect to the natural environment. While constrained by norms against particular nonlethal weapons technologies, states might still make humanitarian arguments for particular technologies to be exempted in the name of saving lives (Fidler, in Lewer, 2002: 28–31).

Indeed, in "The Command of Biotechnology and Merciful Conquest," Guo (2006: 1150–1151) references Clausewitzian logic, arguing that the purpose of war is to break the resistance of the enemy and that this is most humanely and efficiently accomplished with the most precisely targeted attacks possible. At the same time, nonlethal biotechnological warfare is the best option for destruction to "remain civilized":

The goal of precision injury is not necessarily to terminate a life, but to choose a degree of injury depending on the purposes of operations and the types of enemies. By means of gene regulation, certain, or a couple of, key physiological functions in a human body—such as learning, memorizing, balancing, fine manipulation, and even the "bellicose" character—can be injured precisely without a threat of life. . . . After the goal of military operation is achieved or erroneous attack happens, vaccines, drugs, or information about the damaging factor and damaging target can be provided . . . [demonstrating] the greatest mercifulness. Therefore, biotechnology aggressiveness gives rise to relatively merciful conquest as compared to other weapons. . . . This is what is expected by . . . weapon ethics in the 21st century.

Along with certain chemical agents, existing bioweapons could be defined as nonlethal weapons. The US Department of Defense describes nonlethal weapons as "discriminate weapons that are explicitly designed and employed so as to incapacitate personnel or material, while minimizing fatalities and undesired damage to property and environment." The primary object of this category of armament is to target the resolve of the recipients rather than to inflict damage; indeed, recognizable physical damages may not be among the effects of the weapons. Still, it is somewhat problematic to term particular types of weapons as lethal or nonlethal when conventional arms actually kill only about one-quarter of the casualties that they produce and when a small percentage of supposedly nonlethal weapons such as rubber bullets do produce fatalities (Rappert, 2001: 567–568).

Incapacitants, including BW agents, have traditionally been dismissed as ineffective tools of warfare, often because their effects are not immediate and involve too much uncertainty beforehand (Klotz and Sylvester, 2009: 30). These criticisms of the expected lack of utility conferred by BW, however, have been constructed around active combat scenarios in which otherwise conventional and roughly symmetrical forces are confronting each other.

But incapacitants, which include simple biotechnologies like pepper sprays, are also used by domestic law enforcement agencies and by military organizations for crowd control during peacekeeping missions. Critics charge that such devices lead to a greater readiness to use force against civilians, in some instances being used as forms of punishment rather than to prevent greater violence (Rappert, 2001: 563, 568, 575). By making "physical conflict more likely by making it less costly," nonlethal weapons are arguably "contributing to the militarization of police forces and the paramilitarization of militaries" (Rappert, 1999: 741–742). A related concern is that battlefield commanders will, by using nonlethal weapons to clear civilians from areas of operations in the name of their safety, erode restrictions that protect noncombatants from their use. Research indicates that TASERs and pepper spray are now used routinely in situations that did not prompt the use of force before their invention, demonstrating the potential for incapacitants to be used for new purposes of noncombatant control rather than preventing enemy combatant fatalities (Massingham, 2012: 679, 682).

Indeed, after the difficulties it encountered in waging effective urban warfare in Mogadishu in 1993, the Pentagon began intensive research into nonlethal weapons, including colored strobe lighting and synthetically produced odors to nauseate crowds.[7] At approximately the same time, the FBI consulted with counterparts in Moscow about the feasibility of using Soviet technology to broadcast subliminal messages to cult leader David Koresh, who was holding hostages in a protracted standoff with law enforcement, in an attempt to persuade him that the voice of God was ordering him to stand down (Barry and Morganthau, 1994).[8]

By virtue of such advances in nonlethal weapon technologies (including those under development by the US military as described in chapters 2 and 3), scientific developments come to shape not only doctrine but how relations are constituted between actors including soldiers, police, protesters, rioters, insurgents, and bystanders.[9] Power relationships change with the advent of new technologies and also with the ease with which—perhaps bloodless—

violence might be enacted. And with these changes also come new sets of questions concerning the conduct of just war (Rappert, 1999: 746–748).

Historically, the use of nonlethal conventional weapons on the battlefield was regarded as "evil in itself" because they would leave behind maimed victims rather than simply killing. In the Saint Petersburg Declaration of 1868, which occurred between the first Geneva Convention and the Hague Conventions, a commission comprised of representatives of numerous European militaries declared that "the only legitimate object of war should be to weaken the military force of the enemy, which could be sufficiently accomplished by the employment of highly destructive weapons. With that fact established, the delegates agreed to prohibit the use of less deadly explosives that might merely injure the combatants and thereby create prolonged suffering of such combatants." The delegates therefore agreed to ban explosive projectiles weighing less than four hundred grams or any small-arms ammunition bearing incendiaries or reactive chemicals (Krickus, in Walkin, 1986: 420–421).[10]

Some have argued that by opposing the use of nonlethal armaments by military and police forces, humanitarian organizations are actually limiting them to continuing to use only lethal implements rather than causing them to abandon coercive arms altogether. Only with great skill and luck can lethal armaments be used only to incapacitate, and if deliberately fired away from the target they might be used as a signal that the wielder has the weapon and will use it in a lethal manner if further provoked. Of course, using a weapon to signal resolve could also be interpreted to mean that the opponent does not actually have the will to use it or else would have done so the first time, whereas unconventional incapacitants can be used actively while minimizing loss of life (Lewer, 2002: 12–20).

Shifts in whether nonlethal weapons are viewed as either malicious or merciful have as much to do with technological advances—from picric acid to proteomics—as they do with different and evolving constructions of meaning. Rappert (2001: 565) notes that "diametrically opposed interpretations have been offered, for instance, on whether the deployment of such weapons helps to escalate or to minimize conflict." These views are pertinent not only to whether it is permissible to use particular types of advanced BW such as reversible direct-effect weapons but also, and perhaps more challengingly, how to respond to them in kind—not to mention bioterrorist incidents.

BIOLOGICAL ATTACKS AND PROPORTIONAL RESPONSE

One relevant question that must be asked is what is an appropriate proportional response to the use of biological weapons by either state or non-state actors? Unlike a nuclear attack, there are many gradations of potential biological attacks against the United States, and it is not evident that any or all of them justify the type of nuclear retaliation threatened against bioweapon proliferator Iraq in the confrontations of the 1990s and 2000s. Instead, a new host of questions of just war arise with advances in biotechnology.

As the Amerithrax case demonstrates, it may be several years before the perpetrator of a significant biological attack is even identified. Are Cold War–type threats of massive retaliation credible when adversaries are amorphous, or justifiable when a period of years has passed since the occurrence of the attack? Even if a suitable target for retaliation against a non-state actor like al Qaeda could be determined, what would be an appropriate response to an ineffectual use of anthrax against vaccinated military personnel? Could the federal government or the American public justify a WMD response to a WMD attack even if it did not produce massive casualties?

Or what if an agent such as brucellosis, which only incapacitates, were to be used instead? Some speculated in 2001 that the limited anthrax attacks were salami tactics that would break down the taboo against biological attacks by accustoming the global public to their use. The probable failure of future bioterror attacks to provide a casus belli for massive retaliation will erode the deterrent credibility of both nuclear and conventional forces. Also, deterrence will fail if state actors believe they can mask the source of biological attacks, which is easier to do with biological than with other weapons.

Additionally, proliferation will bring engineered pathogens and biotoxins into the possession of non-state actors, against whom threats of massive retaliation are entirely implausible because they operate, usually clandestinely, within the territories of sovereign states. In particular, the use of CBRN weapons against domestic terrorists would be nearly impossible. Shortly after the conclusion of the Cold War, military planners realized that deterrence had faded as a security guarantor and argued that increased protective measures were required instead (Danzig, 1996). The result was the massive new biodefense and biosurveillance programs, described in the previous chapter, that appear to have dovetailed seamlessly with existing BW research.

Examination of the values underlying national security policy making is an essential step in developing a net assessment of the impact of biotechnology on international security. In the century since the advent of WMD, often marked by the use of chlorine gas at Ypres in April, 1915, the norm influencing international laws and most national defense policies has been that unconventional warfare is to be avoided if at all possible. The primary rationales proffered have been fear of uncontrollable escalation and unavoidable indiscriminate civilian casualties.

However, advances in biotechnology are obviating the familiar balance of terror afforded by twentieth-century bacteriological armaments. The overall advantage still remains with the industrially and scientifically advanced major state powers because of their growing capacity to target resources that are not conventional military targets. Yet these developments raise doctrinal questions that require thoughtful consideration.

Unconventional attacks seem likeliest to occur when the attacker's vital interests are unlikely to be threatened in a reciprocal fashion. If a state-sponsored terrorist group used a genetically engineered lethal virus against Americans, it seems unlikely that the United States would respond in kind against the populace of either state sponsors or host states, favoring instead conventional punitive attacks. But public sentiment could begin to shift in favor of in-kind retaliation if biotech attacks continue, particularly if more "humane" responses such as bioregulator disruptions are available. Alternatively, biotech attacks against industrial or agricultural targets appear to offer a low-cost form of coercion and might be more attractive than military interventions (Guillemin, 2005: 7).

It is therefore imperative to set the parameters for acceptable military and counterterrorism usage of advanced biotechnologies prior to any such incidents. While retributive symmetrical attacks are permitted under international agreements on bioweapons, and while states have historically reserved the right to first use of BW if attacked, even extending to allies of the attackers, it is not clear that retributive attacks against civilian populations constitute just war (Guillemin, 2005: 5). For example, if an Iranian-backed non-state group such as Hezbollah used a genetically engineered lethal virus against Americans, could the United States respond in kind against Tehran?

Even an attempted nonlethal response would likely be problematic. As noted, even armaments intended to incapacitate but not to kill rioting or protesting mobs have still killed or seriously wounded hundreds of targets (Rappert, 2005: 214). The major state BW programs also developed "nonlethal" weapons like brucellosis, which causes intense illness over several days but not more than that. Obviously such an outbreak would incapacitate opposing troops and civilian defenders much as the botulants and purgatives used in sieges in the ancient world did, and it would also raise the question of an appropriate response. But given that approximately 2 percent of those infected die from brucellosis, it is difficult to guarantee that such an attack would be truly nonlethal, and the same holds true for the deployment of other emerging biotechnologies with military applications (Guillemin, 2005: 7).[11]

The development of novel technologies to mitigate or even reverse damage to the bodies and psyches of combatants raises a different set of questions. Military research into the treatment of injuries ranging from PTSD to amputation reflect a trend in recent decades that makes it more dangerous to be a civilian in a war zone than a combatant. Not only are local noncombatants unarmed, but wounded soldiers would be priority recipients of what would at least initially be expensive and limited treatments, which would produce the same type of negative public reaction as when stationed government and military personnel receive treatment after terror attacks and local civilians do not.

SECURING RIGHTS AT HOME

Questions of justice and fairness also pertain to defensive measures against domestic biotechnological attacks. Past BW incidents also demonstrate that the effects of attacks and accidents are not distributed uniformly across society. For example, segments of the population with weaker immune systems by virtue of their age are particularly vulnerable (US House of Representatives, 2005: 33). Should biodefense resources be directed first to the most likely targets of infection, as is the influenza vaccine during annual flu season and during potential epidemics? Or to military personnel and those involved in key functions of governance and infrastructure as the Bush administration proposed after 9/11?

The age-old trade-off between liberty and security also takes on new dimensions in the biotech era, in which both the availability and the withholding of information can potentially cause pandemics, both natural and manmade. The ethical application of biotechnologies for the purposes of homeland security and domestic defense therefore requires as much consideration as their potential use on the battlefield.

Equitable Distribution of Resources

Already, if the evidence from BW incidents in urban centers has not indicated that the socioeconomic status of victims plays a role in determining government responsiveness, it at least influences how different segments of the public *believe* they have been treated or would be if faced with an attack. In the Amerithrax incident, with the postal service used as a delivery system for a bioweapon, it was interns and administrative assistants who opened the envelopes, not the political and media stars to whom they were addressed. Nearly half of the deaths and most of the serious illnesses that resulted occurred among minority, working-class USPS employees who did not have the luxury of choosing which federal agency would tend to their health and which would inspect their workspace as did the staff of the US Senate majority leader (Daschle and D'Orso, 2003: 156–159).

By contrast, the *Washington Afro-American* reported that "hundreds of Black postal workers from the now-closed Brentwood mail facility say they fear for their health and likely won't return to work . . . citing widespread mistrust of postal and government officials" as their reasons for retiring or finding other employment (Williams, 2003). Brentwood employees neither believed their supervisors' initial assurances that they were safe nor their later recommendations for monitoring and treatment because of "workplace problems exacerbated by decades of adversarial labor-management problems. Postal workers were suspicious of managers due to labor disputes, and didn't listen to them" (US Government Accountability Office, 2008: 14).

Whether the distrust felt by Brentwood and other postal workers was a contributor or just a product of the same sociological factors, a number of public opinion surveys have demonstrated that African-Americans are more likely to believe that government responses to bioterror attacks would distribute resources less fairly to them than to other demographic groups in the United States. Younger respondents across all ethnic groups believe that race plays a factor in the provision of emergency response, as do those who live in neighborhoods with higher crime rates (Eisenman et al., 2004). African-Americans also have lower levels of trust in both the public health system and community response in general to attend to their needs after a bioterrorist incident (Meredith et al., 2007; SteelFisher et al., 2012; Vaughn et al., 2012).

The direct experience of discrimination by minority groups bears upon views of homeland security failures, not only in the Amerithrax case, but also in conspiratorial views of AIDS or the 2005 rupture of the levees in lower New Orleans. But nearly a quarter of all Americans would expect African-Americans to experience discrimination in the event of an outbreak, and nearly three-quarters believe that wealthy citizens would be the first to receive vaccinations during an outbreak. The possibility therefore exists of race riots in conjunction with biological attacks on urban centers if the situation is particularly bad and the perception holds that access to resources has been particularly unfair. It is also all too easy to imagine mob violence turned against minorities under such circumstances, for instance if terrorized citizens accept the proposition advanced by Maurer and O'Hare (in Maurer, 2009: 451) "that illegal aliens—already fearful of detection and deportation—would hide illnesses, remain untreated, and resist public health officers."

Socioeconomic status clearly plays a role on how biodefense resources are allocated, even if the effect is indirect and based on lower trust in government and willingness to seek treatments. Alibek (1999: 74) was surprised that the victims of the Sverdlovsk anthrax outbreak were mostly adult males, a group that would seem to have among the strongest immune systems and thus be least likely to succumb. He has claimed (although the timing has been disputed by other sources) that the pathogen leak occurred during hours when it was men who were working night shifts and thereby exposed. Mistrust of government among the exposed Soviet population, along with reports of side effects from the vaccine, led to twenty thousand out of the fifty thousand believed to have been in the exposed area to fail to report to receive treatment (Guillemin, 2005: 142–143).

Distrust of government and the power of rumors in panicked mobs was also on display in the Dark Winter planning exercise, conducted just prior to the attacks of 2001. In this simulation of an outbreak of smallpox in the United States and an insufficient number of vaccines available to protect the general population, participants expected to produce riots around facilities that did possess treatments. In the estimation of one participant, former senator Sam Nunn, by the time the three-day simulation had concluded, the problem had shifted from one of logistics to one of preserving democracy when supplies were limited and generally only elites could receive them (Lakoff, in Lakoff and Collier, 2008: 50–52).

Studies of disasters indicate that "trust in government usually rises significantly in the early phases of a crisis but erodes afterward." In part this is because victims and populations that feel threatened turn to authority to protect them, and more so because responders, more from imperfect information than needs for secrecy, provide them with information that is vague or is quickly invalidated by unfolding events. Once lost in these situations, trust in authorities responsible for public safety is difficult to restore (Maurer and O'Hare, in Maurer, 2009: 454). This is as true among elites, such as the congressional staff affected by the Amerithrax attack, as it is among lower socioeconomic status, low-information populations (North et al., 2005).

Amerithrax also demonstrated another difficulty that will confront emergency responders during biological attacks or outbreaks. Despite the fact that fewer than fifty individuals demonstrated sufficient exposure to *Bacillus anthracis* spores to warrant treatment, approximately one hundred thousand residents of the National Capitol Region indicated in surveys that they considered themselves to have been exposed. Biodefense planners can expect large numbers of psychogenic illnesses to be reported, producing strains on resources (Reissman, 2006: 452). Additionally, because stress produces physical strains, terrorized members of the public would experience somatic symptoms including racing heart and upset stomach, leading them to falsely report infections or exposures (Maurer and O'Hare, in Maurer, 2009: 451).

Syndromic Surveillance and Privacy Rights

Such reports matter, not just because of questions of how to care for the healthy who believe that they are ill, but because of how data is collected to alert responders to patterns of outbreaks. Described in the previous chapter, programs like the United States' BioSense alert network rest on the relatively new process of syndromic surveillance: "The ongoing systematic collection, analysis, interpretation, and dissemination of health-related data preceding diagnosis, searching for indicators suggesting sufficient probability of an outbreak to warrant a public health response" (Nordin et al., 2008: 802).

The practice began with the New York City government in 1995 in an effort to monitor ambulance dispatch calls for outbreaks of waterborne illnesses such as giardia, expanded its focus in 1998 to monitor for early warnings of bioterrorism, and expanded again after 9/11 to collect data from all emergency room visits in the city (Heffernan et al., 2004). The program was a response to an outbreak of parasites in the Milwaukee municipal water system in 1993 that went unrecognized until local officials heard a pharmacist announce in a television interview that he had run out of his supply of antidiarrheal medications. Rather than wait for diagnoses of particular diseases, the New York program and those that have followed monitor unusual patterns of symptoms from "nonspecific data sources" such as emergency calls, pharmaceutical sales, emergency room admissions, and nursing home reports. With syndromic surveillance, there are therefore no universal or absolute figures to trigger emergency mobilization. Instead, different agencies have different levels of sensitivity based on their willingness to tolerate false alarms. And because being able to recognize unusual patterns means first having data on normal and seasonal patterns of demand for medical treatment and pharmaceutical purchases, the data must now be collected continuously from the public (Fearnley, in Lakoff and Collier, 2008: 61–68).

In the United States, the national BioSense program receives data from sources including state and regional syndromic surveillance systems, national private health-care corporations, and national laboratories. The data that these entities provide include patient demographic and clinical information and admission and discharge dates (CDC, *BioSense Overview*, 2009). For this reason, "syndromic surveillance may require disclosure of identifiable health data even if the initial transfers that make up the majority of data exchanged involve only non-identifiable data" (Hodge, Fuse Brown, and O'Connell, 2004: 73–80).

As with most national biodefense programs over the past century, syndromic surveillance has attracted little public attention in the United States despite the fact that it is congruent with a number of debates on the appropriate role of the national security state versus individual civil liberties, national versus local control, and patient privacy. However, some local public health workers have criticized BioSense for draining resources away from programs needed for diagnostic health care for what is actually a national security instrument. Federal officials counter that they have been using military resources to treat domestic outbreaks since 1874 and that it is necessary to first collect accurate information that is not always readily forthcoming from state and local health officials. Presaging the BSE and Severe Acute Respiratory Syndrome (SARS) outbreaks of the 1990s and 2000s, the state of California attempted in 1900 to contain economic damages by initially denying the occurrence of an outbreak of plague and rejecting national assistance (Fearnley, in Lakoff and Collier, 2008: 63–65). Given recent discussions in the United States over whether states have the right to opt out of national health-care programs, it seems likely that the debate over federalism will eventually turn to data collection for the purposes of syndromic surveillance.

In 1998, the US Department of Defense launched the Electronic Surveillance System for the Early Notification of Community-Based Epidemics (ESSENCE) program to collect syndromic data on military personnel. This subsequently led to the Global Emerging Infections Surveillance and Response System (GEIS) that monitors troop health worldwide, and ESSENCE II, which collects all hospital data (including for civilians) in the National Capitol Region surrounding Washington, D.C. "Under the military's medical system, every patient encounter produces an electronic record that describes all medical procedures as well as preliminary diagnoses . . . [that] are linked with patient demographic data, including geographic codes for residence and workplace." Initially these data were collected by the Pentagon's new Office of Total Information Awareness, which had come under controversy for domestic electronic surveillance (Fearnley, in Lakoff and Collier, 2008: 70–76).

Beyond questions of federal versus local jurisdiction, there is also the matter of patients' private data being shared between biodefense programs that support the generation of dual-purpose technology. Put another way, if citizens did not want their individual health records to be used in biotechnology research with potential offensive military capabilities, they would apparently not have any say in the matter.

Questions of individual privacy and patient rights also extend to the health-care sector. Syndromic surveillance programs monitor aggregate data, and collected CDC files on individual data within broader surveys do not contain personal identifiers.

> The Health Insurance Portability and Accountability Act (HIPAA) [of 1996] as well as the Common Rule, which promulgates rules for protecting study participants in federally sponsored research programs, provide regulations safeguarding protected health information. . . . Unlike HIPAA, public health law is legislated by individual states rather than by the federal government and has no unifying mechanism for balancing privacy rights against public safety. The proper interaction between and appropriate application of HIPAA, the Common Rule, and public health law during a suspected epidemic is unclear. (Nordin et al., 2008: 802–803)

However, HIPAA does permit the collection of otherwise protected health information by public health authorities for surveillance purposes in the interests of national security, law enforcement, judicial proceedings, or public health purposes including serious threats to public safety. Employee medical data were provided to the US Postal Service after the Amerithrax incident, although in other cases health-care providers have been reluctant to provide records for fear of liability over HIPAA violations. However, "it is inevitable that individually-identifiable health information will be shared with many persons" for their own safety, such as nurses and lab technicians working with select agents. Further potential for identifying individuals exposed in a biological attack would come from records of conspicuous treatment methods or care by biodefense agencies. Currently there are no clearcut guidelines directing what information must be provided to patients about their own conditions and risk levels when doing so might involve a conflict with national security interests. Nor are there regulations in place that would prevent this information from being used by health insurance companies in setting future premiums (Hodge, Fuse Brown, and O'Connell, 2004: 73–80). As illustrated in this book's preface, there are not necessarily rules in place governing patient insurance or worker rights in the event of biological attacks.

Finally, recent advances in an area of biotechnology unrelated to epidemiology will potentially provide the opportunity to severely constrain terrorist and criminal activity, but at a far greater cost to individual privacy than the availability of medical records:

> Research in the field of neurosciences has produced commercial applications in fMRI (functional magnetic resonance imaging) that display portions of the brain that become active when subjects are lying. A number of corporations already employ this technology, but the main clients are spouses investigating infidelity. Beyond this, the implications for homeland security through airport screenings are obvious, as is the potential threat to human rights when both the private and public sectors have access to technology that makes concealing information impossible. (Huang and Kosal, 2008)

Given that imaging technology that monitors body temperature is being deployed in airports throughout the world to prevent the transit of infectious passengers, while other monitors display the passengers' naked bodies to prevent the carrying of concealed devices, it is not difficult to imagine such new types of "lie detectors" becoming routine screening devices. DARPA has already invested in mind-reading technology (Moreno, 2006: 97), and by 2011, researchers at Berkeley publicly demonstrated technology that allowed them to create digital video clips of images reconstructed from subjects' brain activity (Anwar, 2011).

INFORMATION TRANSPARENCY

While brain scans in public facilities were presumably still science-fiction scenarios in the 2010s, a number of other developments had demonstrated that international security concerns do arise when biotechnologies produce data that are not made universally available. Scientific discovery, human security priorities, and national interests have come into conflict as a result of biotechnological advancement, and the dilemma appears to be a structural one. As Guillemin (2005: 200) notes, "the protection of American civilians, the stated biodefense goal [of the United States,] limits beneficence to one population and presumably withholds it from enemies, however broadly or narrowly they are defined."

But the United States is hardly unique in simultaneously striving for biosecurity and securitizing the field of biotechnology in the process. After the terrors of 2001, the first decade of the twenty-first century also witnessed multiple pandemics that posed a challenge for global governance in an increasingly globalized world. Although there were relatively few deaths from outbreaks of SARS in 2003 and H5N1 influenza (avian flu) in 2005, both originating in Asia, and H1N1 influenza (swine flu) in 2009, originating in Latin America, each outbreak required costly coordinated local, national, and international health responses, and international commerce was affected, both in terms of commodities and passenger travel. Local and national economies felt the effects of quarantines, and governments had incentives to limit information.[12]

The World Health Organization's (WHO's) International Health Regulations require "participating states to notify the WHO of a potential 'public health emergency of international concern' so that an epidemic can be contained before it spreads across borders . . . [but] disease surveillance and reporting remains a difficult and demanding task, however, and outbreak information is not always provided by WHO member states on a timely basis" (Commission, 2008: 37). For example, China initially refused to either acknowledge the SARS outbreak or to request international assistance (Enemark, 2007: 29).

One success of global governance, "the influenza tracking system, is currently the best available in disease surveillance—it is a global system and is used annually" (Gronvall et al., 2009: 439). At the regional level, disease surveillance networks are also effective not only from a technical standpoint, but for building shared governance capacity:

> For example, members of the Middle East Consortium on Infectious Disease Surveillance (MECIDS), a regional disease surveillance network of public health experts and ministry of health officials from Israel, the Palestinian Authority, and Jordan, have coordinated the screening, laboratory testing, and risk communication strategies to detect and control 2009 H1N1 influenza. . . . The H5N1 outbreaks in poultry crossed the borders of all 3 MECIDS member countries in under 10 days. However, the surveillance network's strategic planning and uniform response prevented human infection and created public confidence. (Gresham et al., 2009: 399, 401)

But such tools depend on cooperation based on recognition of mutual benefit, and some states have withheld pertinent data in the name of national security, perhaps because they believe that the information will not benefit them directly (Gronvall et al., 2009: 439). For example,

> bucking WHO protocol, Indonesia—a WHO member country—opted to share samples from only 2 of more than 135 people infected with the deadly strain of H5N1, and subsequently ceased providing the WHO with timely notification of outbreaks of the flu in birds and humans. By doing so, it not only openly defied WHO international health regulations and agreements in the name of biodefense and competitiveness, it also claimed "sovereignty" over the viruses that had surfaced within its borders. Since then, openly anti-Western states have helped fuel a global movement further endorsing the notion of "viral sovereignty." (Holbrooke and Garrett, 2008)[13]

The notion that a state can claim sovereignty over a particular virus or microorganism that originated within its borders raises a number of questions, including whether GMOs might therefore be considered the property of a particular government —including the augmented soldiers described in chapter 3. Similarly, after the H1N1 outbreak began there, "Mexico's own government criticized its scientific community for sharing the samples with foreign agencies at all, and promptly invested approximately $330 million in order to build up its own public health infrastructure and enhance the country's capability to deal with threats domestically" (*Cell*, 2010).

One hazard evident from such cases is that governments pursuing what they view to be the national security interests of the state may not understand the security risks posed by their own biological samples or their novel biotechnologies. F. W. de Klerk, the president of South Africa who oversaw the end of apartheid, insisted that he only approved the research of Project Coast (described in chapter 1) because he understood it to be defensive biotechnology (Koblentz, 2009: 133). When a US Department of Defense report argued in 2004 that research into "applications of biological, chemical, or electromagnetic radiation effects on humans should be pursued," it is unclear to what extent the dual-use nature of this research was considered (Klotz and Sylvester, 2009: 19).

As with nuclear technology, once advances are made in biotechnology, it becomes exceedingly difficult to prevent the dissemination of information or efforts to duplicate research. The American Society for Microbiology reacted to the post-9/11 and Amerithrax security environment by announcing in 2003 that it would monitor peer-reviewed submissions and possibly withhold publication of sensitive information from its journals (Guillemin, 2005: 201). Leaving aside the inevitable debate over the virtue of attempting to restrict the progression of science, it is almost certain that such efforts would ultimately be futile in preventing the transmission of data. Ken Alibek (1999: 258–261, 272) came to this conclusion upon finding the data of his former Biopreparat colleagues in international journals, even when those who sponsored them could not have failed to realize their offensive potential, and finding others advertising their knowledge in pathogen manipulation in trade journals. With no evident constraints on information flows, advancing biotechnology continues to lead both researchers and policy makers into uncharted territories of ethics.

NO EASY ANSWERS

The ethics and morality of war in general, and even of specific weapons systems, have been the genesis of heated debate probably since the first act of human combat. But no historical precedent exists for debate about the morality of improved/bioengineered body armor or the ethics of enhanced soldier performance. The bioengineered future of the battlefield steps squarely into the middle of the ongoing debate about genetic engineering and presents policymakers with an unprecedented challenge. (Armstrong and Warner, 2003)

Biotechnology has had an uneasy fit with the warrior ethos since the beginnings of civilization; likewise the Hippocratic oath has generally not lent itself well to human-subject research for military purposes. But in the twenty-first century, biotechnology has become such a pervasive element of both defense and homeland security planning that many questions surrounding its ethical role in international security will not be possible to shunt aside for long. Biosecurity, which relates to the health of individuals and populations, is not necessarily synonymous with national security (Lakoff and Collier, 2008: 8), and there are few guideposts available for navigating some of the new questions emerging with scientific advancements.

Extensive new biodefense institutions have emerged in numerous countries since 2001, and patient rights to privacy typically fall before arguments for national security. The biological threat against which nations must be secured is generally presumed in the twenty-first century to originate with non-state actors, particularly terror networks. Al Qaeda affiliates have thus far refrained from using CBRN weapons in any of their hundreds of attacks in poorly secured Iraq, Afghanistan, or Pakistan. But ISIS has evidently used chemical weapons against local opponents (Associated Press, March 14, 2015), and the so-called Laptop of Doom demonstrates that ISIS is also attracting volunteers with some scientific knowledge and interest in rudimentary BW, so it is reasonable to conclude that more attacks are inevitable.

It is less clear how response resources can be triaged effectively to avoid public mistrust. Considering that most reported illnesses during and after outbreaks are psychosomatic, many individuals who believe that they are afflicted are likely to be dissatisfied with governmental responses that inform them that they are not priorities for treatment. Should apprehended bioterrorists be held liable for the costs of tending to the terrorized who have not actually been exposed? Does a community clinic medical technician on an Indian reservation who has been frightened by the prospect of receiving cross-contaminated mail from the nation's capital have standing to sue for damages? If a local al-Qaeda affiliate or lone ISIS supporter commits an attack, can the entire network be held responsible and subject to punitive attacks abroad, even without evidence that senior leadership was involved?

When bringing military force to bear against either non-state actors or other state militaries, which emerging biotechnologies are consistent with traditions of just war? Many new genomic weapons under development could easily be made nonlethal, but does that mean that they are actually more humane, a view articulated only over the past 150 years of human history? Modern biological weapons were conceived a century ago as terror weapons to demoralize civilian populations, and the newest iterations continue to bear that potential. It is not difficult to imagine a scenario in which everyone's blood pressure in Al-Raqqah suddenly rises to dangerous levels with the promise made by leaflet that conditions will be returned to normal once the location of ISIS leadership is revealed. But would such a move be justifiable?

In the Middle Ages, when poisons were being used in regicidal assassinations, political and religious leaders went to great lengths to portray these bioweapons as uncivilized because they undermined the idea of war and political intrigue as being the "sport of kings" and the prerogative of the powerful. The asymmetric advantage conferred upon poisoners by the available biotechnology of the time undermined the class structure of warfare and prompted calls to respect the chivalry of the "fair fight" that would continue to favor the best-entrenched actors. The Hague Conferences held during the autumn of the Concert of Europe regime were similarly intended to discredit the use of munitions that would confer an asymmetric advantage on the attacker (Price, 1997: 25, 34). Similarly, little outcry is now being raised against the new biotechnologies of warfare by the principal powers of the modern international system, which are developing them and who expect to reap the greatest asymmetric advantages.

In the genomic age, when it may be possible to press a button and cause the adversary's kidneys to fail, military doctrines of *jus in bello* will require new examination. There must also be a recognition that new powers, such as China, that were never a part of the Western Grotian tradition might not share the same normative perspectives on the use of advanced biotechnology on the battlefield. Even seemingly simple questions, such as the responsibilities of states to retired augmented soldiers, will require new doctrines. If an individual has been physically or genetically modified in an irreversible fashion, are their augmentations proprietary technologies of the state?[14] If not, what agency will prevent them from abusing their abilities upon return to civilian status?

Debates over the appropriate roles of various biotechnologies in society and in warfare date back centuries. It is the advent of new genomic rather than pathogenic weapons, and genetically engineered soldiers with biomimetic gear who have tremendous advantages over ordinary opponents, that represent a new challenge for the ethics of warfare. The electronic communications revolution is similarly creating tensions between state security and civil liberties in the area of biodefense. As biotechnologies continue to emerge and progress, these debates will only multiply and intensify throughout the next century of modern biological warfare.

NOTES

1. The international reaction against Syria's use of CW against its own population in 2013 indicates that the prohibition norm is deeply entrenched.

2. The United States did at least debate the development of CW during the Civil War, with one proposal to the secretary of war arguing that projectiles containing liquid chlorine would disable targets as effectively as breaking their limbs. While the author did note the questionable morality of such a strategy, he nonetheless contended that it would humanely "lessen the sanguinary nature of the battlefield" (Christopher, 1994: 203).

3. Definitions of this type of weapon include those that cause superfluous injury and unnecessary suffering, such as specific diseases, abnormal physiological or psychological states, or permanent disfigurement. They also include weapons that produce a field mortality of greater than 25 percent or hospital mortality greater than 5 percent (Rappert, 2005: 217).

4. Grotius also argued that those who used poison to assassinate rulers deserved "fear of disgrace" (Price, 1997: 24).

5. For example, "American soldiers in WWII were known to rub excrement on their bullets in order to cause infections," which Christopher (1994: 205) describes as a violation of *jus in bello* because enemy forces who had been shot were already no longer combatants.

6. Presumably, however, the perpetrators of all forms of warfare believe that they are fighting to ensure future stability and less bloodshed as a result, which would seem to make the double effect a sincere argument for exoneration by any and all belligerents.

7. Lewer (2002: 3) argues that the use of malodorants, produced from toxins and other biological material, in riot control demonstrates their potential for battlefield operations but potentially offers a slippery slope toward the use of other biological agents.

8. Although this novel approach was apparently not taken and the standoff ended with the deaths of dozens of unarmed cult members, the situation could have been worse: nine years later, Russian efforts to end a hostage standoff at a Moscow theater that relied on fentanyl, a supposedly nonlethal knockout gas, killed over one hundred of the hostages, whereas the terrorists up to that point had killed only one. The Russian familiarity with the gas lends credence to reports of calmative agents used during the Soviet invasion of Afghanistan (Moreno, 2006: 140, 144). As Dando (in Danzig and Tucker, 2012: 167) notes, "a study for the European Defense Agency in 2007 suggested that 'calmative' drugs could be used to clear facilities, structures, and areas, indicating that a normative threshold may have already been crossed."

9. Klotz and Sylvester (2009: 29) argue that incapacitants used for domestic law enforcement purposes are also illegal under the BWC. Under the terms of the convention, however, select agents may be retained for "protective or peaceful purposes," which could be argued to include force protection and crowd control.

10. Price (1997: 34) describes the Hague regime as a declaration of self-restraint by the strong, one intended to delegitimize the use of BW by weaker actors seeking to level the playing field. In this regard, it was very much like the BWC seventy years later.

11. Likewise, the 2002 biodefense debate over mass smallpox vaccinations in the United States was ultimately determined by the expected deaths of a small percentage of recipients.

12. In June 2003, I walked through the utterly deserted international terminal at Toronto's Pearson Airport on the eve of a summer holiday weekend when that city was the only metropolis outside of Asia to experience a SARS outbreak (via its sizable Chinese émigré community). In May 2009, I was subjected to an infrared scan while attempting to disembark an airplane and enter Hungary during the height of swine flu fever, and the most pressing security question that the immigration officer at Israel's Ben Gurion Airport had for me three days later was whether I had been to Mexico recently.

13. In 2011, Indonesia and the WHO agreed to create the Pandemic Influenza Preparedness Network, which provided guidelines permitting sharing of both viruses and the profits derived from them (Garrett, 2012). Davies, Kamradt-Scott, and Rushton (2015) argue that Jakarta's developing position is the result of evolving international norms of cooperation on biosecurity as a global governance concern.

14. Miranda et al. (2015: 64) note that DARPA neural implants should be designed to last for decades and accept continuous upgrades to avoid the need for repeated surgeries.

Conclusion

The Modern Prometheus at War

A race of devils would be propagated on the earth who might make the very existence of the species of man a condition precarious and full of terror. Had I a right, for my own benefit, to inflict this curse upon everlasting generations?
—Mary Shelley, *Frankenstein; or, The Modern Prometheus* (1818)

While concerns about the impact of biotechnology on international security appeared in works of fiction two hundred years ago, the accelerated development of the field in recent years has ensured that the global community will confront the strategic and ethical challenges that it poses in the near future. Most of the attention in the intervening two centuries paid to the potential applications of biotechnology in world politics has focused on "biological warfare" through germs and viruses that great powers have employed since the 1700s and which have now drawn the attention of non-state actors as well. In this view, the probable proliferation of biological weapons (often called the poor man's nuclear bomb) to terrorists and rogue states is expected to result in threats to conventionally stronger states and greater instability in the international system.

However, the parameters of biotechnology, which are being continuously expanded by private-sector firms as well as governmental research programs, are far broader than the conventionally imagined germ warfare and will actually reinforce the superior security positions of leading advanced industrial states. States with the resources to do so are already integrating into their force structures biotechnological advances—some contemporary, some already decades old—that reduce the effectiveness of attacks against their personnel and equipment or that can destroy enemy resources without firing a shot. The fact that many of these developments are being pioneered by contracted private firms increases the threat of biological weapons proliferation through the black market and even the use of such technologies by private actors against other NGOs. Ultimately, though, the accretion of biotechnologies ranging from advanced polymers and enzymes to synthetic organs enhanced with recombinant DNA will decisively shift the advantage back to established state powers.

INTERNATIONAL RELATIONS IN THE BIOTECH AGE

Many recent analyses of the impact of trends in biotechnology conclude that the United States and other leading developed nations will experience a reversal of fortunes as the technology they developed proliferates to rogue states and terrorist groups. Koblentz (2009: 21) argues that knowledge proliferation means that the capacity to wage biological warfare is extending even to private individuals and that greater international instability will result. The technology of biological warfare, in this view, is weighted toward offense and does not provide an effective strategic deterrent, and practical constraints on development and usage are eroding. Hope for the prevention of biological warfare lies in the fact that genetic science is sufficiently advanced that pathogens can be identified quickly and defenses mounted.

These interpretations, focused almost exclusively on the historical development of pathogens in classical state biological weapons programs, miss the mark on the likeliest impacts of biotechnology on international security. The first lies in the continued applicability of the logic of deterrence. The ability to decode genomes carries not only technical but political implications as well because it is now possible to identify the sources of engineered pathogens. Fears that the 2001 Amerithrax attacks were the work of al Qaeda, encouraged in the letters sent by the perpetrator, faded quickly once it was established that the anthrax spores were from the domestic Ames strain used in research by the US military. Although it was not initially possible based on this information to identify one culprit among thousands with access to the samples, this changed with scientific breakthroughs made during the course of the investigation. The case illustrated that perpetrators can be identified and interventions can be undertaken by either law enforcement or the military. The availability of genetic fingerprinting means that deterrence through the threat of retaliation remains a possibility even in the face of anonymous attacks.

However, while the logic of deterrence remains intact, the new dilemma that emerges is the question of what constitutes an appropriate response to an attack using biotechnology. The use of transmissible pathogens, such as plague or Ebola, or of highly lethal infectious agents such as anthrax, to produce mass casualties is generally accepted as employment of a WMD. But what about attacks that produce only disruptions? Are in-kind responses justified? Is overwhelming force appropriate, and against which targets? These ethical questions must ultimately be addressed by governments and the citizens who empower them.

In their extensive history of the Soviet BW program, Leitenberg and Zilinskas (2012: xiii) argue that Soviet leaders were unethical because they spent prodigious amounts of resources on advanced biotechnological weapons while their impoverished citizens lacked adequate access to education and health care. But it is all too easy to imagine the response of a clinic worker on the Pine Ridge Indian Reservation to Pentagon programs to create gecko-powered super soldiers. There is always a trade-off between guns and butter, between investing in national security and investing in human security. What would Captain America do?

RETRENCHMENT RATHER THAN REVOLUTION

The second major consequence of the full range of biotechnological advancement is that the most technologically advanced actors, already the most powerful in the international system, will gain even greater military edges over their rivals and non-state actors. The application of biotechnology to warfare is not a new concept, but the successful integration of modern molecular biology and biotechnology with conventional power projection capabilities is a development of the RMA that is only beginning to be recognized.

The technologies that make the most sophisticated uses of biotech, including soldier enhancement and direct-effect weaponry, will only be within the reach of the most advanced state actors. Rather than being the "poor man's nuclear weapon," twenty-first-century biotechnology will actually provide a decided asymmetrical advantage to major powers that will complement their superiority in conventional forces. Technologically advanced states will be far more likely to be able to counter classical "germ warfare" like anthrax attacks by rogue states and non-state groups than will be actors bereft of a biotech industry to mount defenses against vectors that introduce terminators or proteomic weapons that disrupt human bioregulators.

What these biotechnologies do not disrupt is order within the international system. In the past, advances in weapons technology have been condemned as immoral in part because the most powerful actors, whether states or rulers, viewed them as challenges to their hegemony. Today, terrorists and rogue states are imputed to have a desire to use bioweapons, meaning to release pathogens against civilian targets, but few outcries have been heard over the legitimacy of the advantages conferred by other biotechnologies upon what are already the strongest actors.

DARPA director Regina E. Dugan began and ended her 2012 TED talk by posing the question, "What would you attempt to do if you knew you could not fail?" At the conclusion, the visual accompaniment to her presentation was an image of a child dressed in a superhero costume that was intended to reinforce her exhortation to the audience to dream as big as they did when they were children. But, in the context of the research she supervised, both the visual and the question carry a different import.

Beyond military dominance, American economic competitive advantages in biotech will persist for some time, although competition is increasing. In particular, China and Singapore both began to invest heavily in biotech in the 1980s. But all biotech powers can be expected to face at least temporarily rising public resistance to the production of GMOs as engineered organisms proliferate and some inevitably escape labs or produce otherwise unpleasant unintended consequences (Chase-Dunn and Reifer, 2002: 1, 7).

One factor complicating this scenario is the role of the academic and private sectors in developing many of these technologies. As noted, many of these institutions have demonstrated that they do not maintain effective security of their facilities or inventories, and transnational corporations may have employees with loyalties to rogue states or non-state external groups. Additionally, some companies may themselves be fronts for clandestine activity by states or paramilitary groups. The possibility of proliferation of advanced technologies through the private sector can therefore not be ignored.

Before accepting these extrapolations, however, it is worth questioning just how significant the commitment to twenty-first-century biowarfare R&D the hegemons are making actually is. Some of the projects or doctrines still described as aspirations by the United States date back to the 1990s and may never be realized. It is unknown how much of DARPA's $2 billion budget goes to its new Biological Sciences Division and to partner universities, let alone to which projects beyond those it announces. Chinese military publications have described an interest in direct-effect weapons and other offensive capabilities, but there is no publicly available information about whether or how it is pursuing them, and the same is true of other potential great powers such as India, Russia, and the leading states of the European Union.

However, the United States has regularly announced spending billions of dollars on military biotech and homeland biosecurity since the Daschle incident in 2001, and it is worthwhile to remember the observation attributed to another US Senate leader, Everett Dirksen: "A billion dollars here, a billion dollars there, soon you're talking real money!" Even if not all of these technologies are developed, the interest expressed and resources invested in exploring them prompts other states to examine developing their own. For example, expressions of interest by China in insect-delivered vectors prompt the United States and India to explore countermeasures, which also necessitates mastering offense-based technology. In this regard, little has changed since the security dilemma drove the development of BW arsenals by the great powers one hundred years ago.

LEARNING TO LIVE WITH SCIENCE FICTION

The ambiguities that surround the proliferation of biotechnologies for security purposes are a cause for viewing their advent skeptically. At the same time, this examination of them is not intended to express an antitechnology Luddite view or, in the words of Kurt Vonnegut, "an anti-glacier book" protesting something inescapable. But the law of unintended consequences seems to impact nearly all scientific developments and many military doctrines, and the author personally experienced one significant unintended consequence stemming from decades of bioweapon and biodefense research.

But while those events might be relegated to a single curious historical case, so many other developments are transpiring that all of us, perhaps the entire human race, must begin to be mindful of the consequences that are intended by decision makers as well as the potential unintended effects that may reshape the twenty-first-century world. Given that rapid advancements in biotech with security applications, or at least implications, are occurring largely beneath the public radar, it is imperative that both policy makers and publics begin to establish parameters for the acceptable use of biotech in warfare—and in peacetime—rather than allow events to overtake them.

New developments in biotechnology will afford the United States, and likely China and other major state actors, with a decisive advantage in power projection. However, other biotechnological developments will destabilize the international system to a degree as some states find their economic output outdated while others assume even greater production capacity. It is incumbent on policy makers to prepare carefully today for a tomorrow that is rapidly approaching.

The developments described in this book are occurring without an informed public debate, and, indeed, many of the biotechnologies I have outlined doubtless seem too fantastical to warrant serious consideration. But just as most decision makers and the general public would have dismissed the plausibility of atomic weapons before Hiroshima and were unaware before the invasion of Afghanistan that unmanned aerial vehicles were already in existence, so too are the seemingly farfetched developments of advanced biotechnology already manifesting themselves in force planning and budgeting in the United States and elsewhere.

Director Dugan of DARPA noted that, at the turn of the twentieth century, experts proclaimed powered heavier-than-air flight a technology of the distant future. Ferdinand Foch subsequently dismissed airplanes as "interesting toys of no military value" a year after the Wright brothers' flight and a decade before he became the highest-ranking Allied commander in World War I. With an eye to her own agency's past, Dugan noted that the prototype Internet that it created in 1969 crashed two letters into the first log-in attempt (Dugan, 2012).

Sir Arthur C. Clarke, noted both for his science-fiction tales and for ushering in the global communications era by devising satellite broadcasting, was fond of remarking that "any sufficiently advanced technology is indistinguishable from magic."[1] Although ethical considerations about the impact of biotech and its possible effects on the stability of the international community were raised in *Frankenstein* two hundred years ago, in the twenty-first century these technologies have become so sufficiently advanced that policy makers can no longer dismiss them as magic. There are now thousands of modern Prometheuses, some in the employ of states, some working for private companies or in academic laboratories, laboring assiduously to steal the secrets of the universe from Olympus.

Many of their discoveries have already been assimilated into both commercial and military equipment. These and others still in development—or not yet publicly announced—will have a profound influence on the maintenance of international security in the twenty-first century. The discoveries are the legacy of centuries of biotechnologies applied for security purposes, and some results are as unpredictable as the role of one such Prometheus in the form of anthrax sent in the mail. However, despite the uncertainties, the ultimate beneficiaries of the biotech revolution will almost certainly not be rogue actors but states with leading biotech sectors, and it is incumbent upon these titans to enact sound policies in their manipulation of the building blocks of life.

NOTE

1. Ajay Lele (2009: 1) also refers to "Clarke's Third Law" in his book on emergent military technologies.

Bibliography

Ackerman, Gary. (2005) "WMD Terrorism Research: Whereto from Here?" *International Studies Review* 7 (1): 140–143.

Adams, Thomas K. (Autumn 1998) "Radical Destabilizing Effects of New Technologies." *Parameters* 28 (3). http://strategicstudiesinstitute.army.mil/pubs/parameters/Articles/98autumn/adams.htm.

Alder, Jeff R. (March 20, 2009) "Human Weapon System Maintenance and Optimization." *Commander's Corner*, Kirtland Air Force Base. http://www.kirtland.af.mil/news/story.asp?id=123140746.

Alibek, Ken. (1999) *Biohazard: The Chilling True Story of the Largest Covert Biological Weapons Program in the World—Told from the Inside by the Man Who Ran It.* New York: Random House.

All China Biotech Conference and Exhibition. (2014) http://en.biotechchina-nj.com/index.html.

Allen, David. (April 14, 2003) "Clotting Agents Buy Wounded Troops Life-Saving Time." *Stars and Stripes.*

Allison, Graham. (1971) *Essence of Decision: Explaining the Cuban Missile Crisis.* Boston: Little, Brown.

Anderson, Andrea. (August 31, 2007) "Chimera Controversy." *Scienceline.* http://www.scienceline.org/2007/08/31/bio_anderson_chimera.

Anthony, Sebastian. (October 4, 2013) "Meet DARPA's WildCat: A Free-Running Quadruped Robot That Will Soon Reach 50 mph over Rough Terrain." *Extreme Tech.* http://www.extremetech.com/extreme/168008-meet-darpas-wildcat-a-free-running-quadruped-robot-that-will-soon-reach-50-mph-over-rough-terrain.

Anwar, Yasmin. (September 22, 2011) "Scientists Use Brain Imaging to Reveal the Movies in Our Mind." UC Berkeley News Center. http://newscenter.berkeley.edu/2011/09/22/brain-movies.

Armstrong, Robert E., and Jerry B. Warner. (March 2003) "Biology and the Battlefield." *Defense Horizons*, no. 25. National Defense University, Washington, DC.

Asian Age. (August 3, 1998) "Seedy Business."

Associated Press. (November 6, 1998) "Steps Taken toward Growing Organs."

———. (June 8, 2013) "Shannon Richardson, Pregnant Actress, Tried to Frame Estranged Husband for Ricin Letters Sent to Barack Obama, Michael Bloomberg, FBI Says."

———. (March 14, 2015) "Islamic State Used Chemical Weapons against Peshmerga, Kurds Say."

Barry, John, and Tom Morganthau. (February 7, 1994) "Soon, 'Set Phasers to Stun.'" *Newsweek.* http://www.newsweek.com/1994/02/07/soon-phasers-on-stun.html.

Basken, Paul. (October 7, 2013) "Defense Research Agency Hunts for Biotech Innovators." *Chronicle of Higher Education.*

BBC News. (March 10, 2009) "Nano-Treatment to Torpedo Cancer."

Ben Ouagrham-Gormley, Sonia. (2014) *Barriers to Bioweapons: The Challenges of Expertise and Organization for Weapons Development.* Ithaca, NY: Cornell University Press.

Benjamin, Mark. (March 25, 2012) "Robert Bales Charged: Military Works to Limit Malaria Drug in Midst of Afghanistan Massacre." *Huffington Post.*

Bergen, Peter L. (2006) *The Osama bin Laden I Know.* New York: Free Press.

Bevan-Jones, Robert. (2009) *Poisonous Plants: A Cultural and Social History.* Oxford, UK: Windgather Press.

Biesecker, Calvin. (July 13, 2008) "DHS Remains on Track for Biowatch 'Fly-off' Next Year." *Defense Daily.*

———. (February 27, 2009) "Obama Proposing $42.7 Billion for DHS, Including $355 Million for Cyber Security." *Defense Daily.*

———. (June 30, 2009) "House Bill Slams DHS BioWatch Program." *Defense Daily.*

Bork, Kristian H., et al. (2007) "Biosecurity Scandinavia." *Biosecurity and Bioterrorism: Biodefense Strategy, Practice, and Science* 5 (1): 62–71.

Brackett, D. W. (1996) *Holy Terror: Armageddon in Tokyo.* New York: Weatherhill.

Breitenbach, Dagmar. (September 9, 2015) "A Fresh Light on the Nazis' Wartime Drug Addiction." *Deutsche Welle.* http://www.dw.com/en/a-fresh-light-on-the-nazis-wartime-drug-addiction/a-18703678.

Bull, Hedley. (2002) *The Anarchical Society.* 3rd ed. New York: Columbia University Press.

Bush, George W. (January 29, 2002) State of the Union Address. http://georgewbush-whitehouse.archives.gov/news/releases/2002/01/20020129-11.html.

Buzan, Barry, Ole Waever, and Jaap de Wilde. (1998) *Security: A New Framework for Analysis.* Boulder, CO: Lynne Rienner.

Callaway, Ewen. (October 1, 2009) "Free-Flying Cyborg Insects Steered from a Distance." *New Scientist.* http://www.newscientist.com/article/dn17895-freeflying-cyborg-insects-steered-from-a-distance.html.

Caplan, Arthur L., and Glenn McGee. (June 7, 2004) "Bioethics for Beginners: An Introduction to Bioethics." *American Journal of Bioethics.* http://www.bioethics.net/bioethics-resources/bioethics-glossary/introduction.

Carr, Edward Hallett. (1939). *The Twenty Years Crisis.* New York: Harper Perennial.

Carter, Ashton B. (2001) *Keeping the Edge: Managing Defense for the Future.* Cambridge, MA: MIT Press.

Carus, Seth W. (1997) "The Threat of Bioterrorism." *Strategic Forum,* Institute for National Strategic Studies.

Cavallaro, Gina. (June 24, 2010) "Standardized Tourniquet, New Bandages for IFAK." *Marine Corps Times.*

CBWInfo. (2005) "Ancient Times to the 19th Century," in *A Brief History of Chemical, Biological and Radiological Weapons.* http://www.CBWInfo.com/History/History.html.

Cell. (January 15, 2010) "Mexican Scientists Reflect on Swine Flu Lessons." SciDev.net. http://www.scidev.net/en/features/mexican-scientists-reflect-on-swine-flu-lessons.html.

Centers for Disease Control and Prevention (CDC). (2016) *National Syndromic Surveillance Program.* http://www.cdc.gov/nssp/overview.html.

———. (2009) *About BioSense.* http://www.cdc.gov/nssp/biosense/index.html.

———. (August 26, 2005) *Morbidity and Mortality Weekly Report.* "Syndromic Surveillance: Reports from a National Conference, 2004." http://www.cdc.gov/mmwr/pdf/wk/mm54su01.pdf.

Chamberlain, Casey. (September 19, 2006) "My Anthrax Survivor's Story." *9/11: Five Years Later.* http://www.msnbc.msn.com/id/14785359.

Chandler, David. (April 12, 2010) "Viruses Harnessed to Split Water." *MIT News.* http://newsoffice.mit.edu/2010/belcher-water-0412.

Chandrasekaran, Rajiv. (March 29, 2014) "A Legacy of Pain and Pride." *Washington Post.*

Chase-Dunn, Chris, and Thomas Reifer. (June 26, 2002) *U.S. Hegemony and Biotechnology: The Geopolitics of New Lead Technology.* Institute for Research on World Systems Working Paper No. 9, University of California–Riverside.

Christopher, Paul. (1994) *The Ethics of War and Peace.* Englewood Cliffs, NJ: Prentice Hall.

Clunan, Anne L., Peter R. Lavoy, and Susan B. Martin. (2008) *Terrorism, War, or Disease? Unraveling the Use of Biological Weapons.* Palo Alto, CA: Stanford University Press.

Coghlan, Andy. (February 14, 2005) "Gene Therapy Is First Deafness 'Cure.'" *New Scientist.* http://www.newscientist.com/article/dn7003-gene-therapy-is-first-deafness-cure.html.

———. (March 13, 2013) "Craig Venter Close to Creating Synthetic Life." *New Scientist.* http://www.newscientist.com/article/dn23266-craig-venter-close-to-creating-synthetic-life.html#.U5Zxj_m1bYg.

Cohen, Eliot. (1998) "Made to Measure: New Materials for the 21st Century," *Asia Pacific* 29 (426).

Cohen, John. (July–August 2002) "Designer Bugs." *Atlantic Monthly.* http://www.theatlantic.com/issues/2002/07/cohen-j.htm.

Cohen, Roger. (February 8, 1998) "The Weapon Too Terrible for the Parade of Horribles." *New York Times.*

Commission on the Prevention of WMD Proliferation and Terrorism. (2008) *World at Risk.* New York: Vintage.

Congressional Research Service. (March 12, 2007) *Agroterrorism: Threats and Preparedness.* Washington, DC: Government Printing Office.

Cooper, Simon. (October 1, 2009) "North Korea's Biochemical Threat." *Popular Mechanics.* http://www.popularmechanics.com/technology/military/4208958.

Dando, Malcolm. (2006) *Bioterror and Biowarfare: A Beginner's Guide.* Oxford, UK: Oneworld Publications.

Danzig, Richard. (1996) "Biological Warfare: A Nation at Risk—A Time to Act." *Strategic Forum,* No. 58–59, Institute for National Strategic Studies.

Danzig, Richard, and Jonathan B. Tucker, eds. (2012) *Innovation, Dual Use, and Security: Managing the Risks of Emerging Biological and Chemical Technologies.* Cambridge, MA: MIT Press.

Daschle, Tom, and Michael D'Orso. (2003) *Like No Other Time: The 107th Congress and the Two Years That Changed America Forever.* New York: Crown.

Davies, Sara E., Adam Kamradt-Scott, and Simon Rushton. (2015) *Disease Diplomacy: International Norms and Global Health Security.* Baltimore, MD: Johns Hopkins University Press.

De Lange, Catherine. (July 9, 2010) "Nanoparticle Bandages Could Detect and Treat Infection." *New Scientist.*

Defense Advanced Research Projects Agency (DARPA). Various articles accessed 2010–2015. http://www.darpa.mil.

———. (June 5, 2014) "DARPA Z-Man Program Demonstrates Human Climbing Like Geckos." http://www.darpa.mil/news-events/2014-06-05.

———. (April 1, 2014) "DARPA Launches Biological Technologies Office." http://www.darpa.mil/news-events/2014-04-01.

Defense Daily. (July 11, 2008) "BioShield Still Has $4 Billion to Spend." http://www.defensedaily.com/publications/dd/BioShield-Still-Has-$4-Billion-To-Spend_3274.html.

Defense Research & Development Organization. (2015) http://www.drdo.gov.in/drdo/English/index.jsp?pg=homebody.jsp.

Department of Biotechnology, Government of India. (2013) *Biotechnology Landscape in India.* New Delhi: FICCI. https://www.tekes.fi/globalassets/julkaisut/biotechnology_landscape_in_india.pdf.

Dewar, Elaine. (March 20, 2015) "SMARTS Update: How the US Military Will Mold Minds with Stories: DARPA's Narrative Networks." http://elainedewar.blogspot.ca/2015/03/smartsupdatemarch202015howus.html.

Dixon, Alex, and Julia Henning. (July 22, 2013) "Nett Warrior Gets New End-User Device." United States Army. http://www.army.mil/article/107811.

Doornboss, Harald, and Jenan Moussa. (August 29, 2014) "Found: The Islamic State's Terror Laptop of Doom." *Foreign Policy*.

Drell, S. D., A. D. Sofaer, and G. D. Wilson, eds. (1999) *The New Terror: Facing the Threat of Biological and Chemical Weapons*. Palo Alto, CA: Hoover Institution Press.

Dugan, Regina E. (March 2012) "From Mach 20 Glider to Hummingbird Drone." TED talk. https://www.ted.com/speakers/regina_dugan.

Duncan, David Ewing. (November 3, 2012) "How Science Can Build a Better You." *New York Times*.

Dyer, Owen. (September 23, 2006) "British Soldiers Are 'Guinea Pigs' for New Use of Blood Clotting Agent." *British Medical Journal* 333 (7569): 618. http://www.ncbi.nlm.nih.gov/pmc/articles/PMC1570861.

The Economist. (November 21, 2009) "The Parable of the Sower."

———. (February 25, 2010) "Handling the Cornucopia."

———. (April 3, 2010) "We All Want to Change the World."

———. (May 22, 2010) "And Man Made Life."

———. (May 22, 2010) "Genesis Redux."

Egudo, Margaret. (2004) *Overview of Biotechnology Futures: Possible Applications to Land Force Development*. Edinburgh, South Australia: DSTO Science Systems Library.

Eisenman, D. P., et al. (2004) "Will Public Health's Response to Terrorism Be Fair? Racial/Ethnic Variations in Perceived Fairness during a Bioterrorist Event." *Biosecurity and Bioterrorism* 2 (3): 146–156.

Enemark, Christian. (2007) *Disease and Security: Natural Plagues and Biological Weapons in East Asia*. London: Routledge.

European Commission. (2013) "A Definition of Biotechnology." http://ec.europa.eu/growth/sectors/biotechnology/bio-based-products/index_en.htm.

Falkenrath, Richard A., Robert D. Newman, and Bradley A. Thayer. (1998) *America's Achilles' Heel: Nuclear, Biological, and Chemical Terrorism and Covert Attack*. Cambridge, MA: MIT Press.

Federation of American Scientists. (2010) "Biosafety Levels Information." http://www.fas.org/programs/bio/resource/biosafetylevels.html.

Fletcher, Amy L., and Christopher S. Allen. (2007) *BioBricks or BioConflicts? Building Public Trust in European Governance of Synthetic Biology*. Paper prepared for the Annual Meeting of the American Political Science Association.

Fox, Jeffrey. (1997) "U.S. Military Takes Up Phytoremediation." *Nature* 15 (612). http://www.nature.com/nbt/journal/v15/n7/full/nbt0797-612a.html.

Franco, Crystal, and Nidhi Bouri. (2010) "Environmental Decontamination Following a Large-Scale Bioterrorism Attack: Federal Progress and Remaining Gaps." *Biosecurity and Bioterrorism: Biodefense Strategy, Practice, and Science* 8 (2): 107–117.

Frank, Thomas. (July 24, 2007) "Non-Combat Deaths in Iraq Drop." *USA Today*.

Fukuyama, Francis. (September–October 2004) "Transhumanism." *Foreign Policy*, 42–43.

Garrett, Laurie. (January 5, 2012) "Flu Season." *Foreign Policy*. http://www.foreignpolicy.com/articles/2012/01/05/flu_season.

Gartner, Scott Sigmund, and Gary M. Segura. (1998) "War, Casualties, and Public Opinion." *Journal of Conflict Resolution* 42 (3): 278–300.

Geissler, Erhard, and John Ellis van Courtland Moon, eds. (1999) *Biological and Toxin Weapons: Research, Development and Use from the Middle Ages to 1945*. Stockholm: SIPRI, Oxford University Press.

Geron Inc. (June 7, 2010) "Geron Presents Clinical Data on Its Telomerase Inhibitor Drug in Breast Cancer at ASCO." Press release. http://www.geron.com.

Geron Press. (April 21, 1998) *In Vivo Data from Telomerase Knock Out Mice*.

Gerstein, Daniel M. (2009) *Bioterror in the 21st Century: Emerging Threats in a New Global Environment*. Washington, DC: Naval Institute Press.

———. (2013) *National Security and Arms Control in the Age of Biotechnology*. Lanham, MD: Rowman & Littlefield.

Gillert, Douglas. (November 9, 1998) "Navy Researchers Test Anti-Malaria Vaccine." American Forces Press Service. http://archive.defense.gov/news/newsarticle.aspx?id= 41776.

Goldblatt, Michael. (2002) *Office Overview.* http://archive.darpa.mil/DARPATech2002/ presentations/dso_pdf/speeches/GOLDBLAT.pdf.

Grafton, Scott T., and Michael B. Miller (2013) "Cognitive Neuroscience." Institute for Collaborative Biotechnologies. https://www.icb.ucsb.edu/research/cognitive-neuroscience.

Gresham, Louis, et al. (2009) "Trust across Borders: Responding to 2009 H1N1 Influenza in the Middle East." *Bioterrorism: Biodefense Strategy, Practice, and Science* 7 (4): 399–404.

Gronvall, Gigi Kwik, et al. (2009) "Prevention of Biothreats: A Look Ahead." *Biosecurity and Bioterrorism: Biodefense Strategy, Practice, and Science* 7 (4): 433–442.

Guillemin, Jeanne. (2005) *Biological Weapons: From the Invention of State-Sponsored Programs to Contemporary Bioterrorism.* New York: Columbia University Press.

———. (2011) *American Anthrax.* New York: Henry Holt.

Guo, Jinglin. (January 1996) "New Development of Military Medical Sciences and China's Policy." *Beijing Renmin Junyi* [People's Military Surgeon]. http://wnc.fedworld.gov.

Guo, Ji-Wei. (2006) "The Command of Biotechnology and Merciful Conquest." *Military Medicine* 171 (11): 1150–1154.

Guo, Ji-Wei, and Xue-sen Yang. (July–August 2005) "Ultramicro, Nonlethal, and Reversible: Looking Ahead to Military Biotechnology." US Army Professional Writing Collection, *Military Review* 85 (4).

The Hague Peace Conference. (July 29, 1899) *Declaration on the Use of Projectiles the Object of Which Is the Diffusion of Asphyxiating or Deleterious Gases.* http://avalon.law.yale.edu/ 19th_century/dec99-02.asp.

Harvard School of Public Health. (November 8, 2001) "Survey Shows Americans Not Panicking over Anthrax, but Starting to Take Steps to Protect Themselves Against Possible Bioterrorist Attacks." http://archive.sph.harvard.edu/press-releases/archives/2001-releases/ press11082001.html.

Heffernan, Richard, et al. (September 24, 2004) "New York City Syndromic Surveillance Systems." *Morbidity and Mortality Weekly Report* 53:25–27.

Hespanha, João P., and Kevin W. Plaxco. (2013) "Control and Dynamical Systems." Institute for Collaborative Biotechnologies. https://www.icb.ucsb.edu/research/control-and-dynamical-systems.

Hodge, James G., Erin Fuse Brown, and Jessica P. O'Connell. (June 2004) "The HIPAA Privacy Rule and Bioterrorism Planning, Prevention, and Response." *Biosecurity and Bioterrorism: Biodefense Strategy, Practice, and Science* 2 (2): 73–80.

Holbrooke, Richard, and Laurie Garrett. (August 10, 2008) "'Sovereignty' That Risks Global Health." *Washington Post.*

Holley, Peter. (November 20, 2015) "The Tiny Pill Making Fighters in Syria's War Feel Like Superhuman Soldiers." *Washington Post.*

Homer-Dixon, Thomas. (January–February 2002) "Weapons of Mass Disruption." *Foreign Policy*, 52–63.

Huang, Jonathan Y., and Margaret E. Kosal. (June 20, 2008) "The Security Impact of the Neurosciences." *Bulletin of the Atomic Scientists.* http://thebulletin.org/security-impact-neurosciences.

Industrial College of the Armed Forces. (2005) *Industry Studies: Biotechnology.* Washington, DC.

Industry Association Synthetic Biology. (2008) *Workshop: Technical Solutions for Biosecurity in Synthetic Biology.*

Ingham, Richard. (January 24, 2013) "Breakthrough in Storing 700 Terabytes of Data in 1 Gram of DNA." *The Age.*

International Committee of the Red Cross. (September 25, 2002) *Appeal on Biotechnology, Weapons, and Humanity.* https://www.icrc.org/eng/assets/files/other/irrc_848_ biotechnology.pdf.

Israelachvili, Jacob N., Kimberly L. Turner, and Katie Byl. (2013) "Design Principles and Strategies for Biomimetic, Gecko-Like Ambulation." Institute for Collaborative Biotechnologies. https://www.icb.ucsb.edu/research/control-and-dynamical-systems/design-principles-and-strategies-biomimetic-gecko-ambulation.

Joslin, Jeffrey. (February 19, 2015) "DARPA Hopes to Give US Soldiers Terminator-Like Vision." *Tech Gen Mag.* http://techgenmag.com/2015/02/19/darpa-hopes-to-give-us-soldiers-terminator-like-vision.

Judson, Olivia. (May 27, 2010) "Baby Steps to New Life-Forms." *New York Times.*

Kahan, Jerome K. (April 1996) "Regional Deterrence Strategies for New Proliferation Threats." *Strategic Forum* no. 70.

Kahn, Jennifer. (November 9, 2015). "The Crispr Quandary." *New York Times.*

Kaiser, Jocelyn. (September 12, 2011) "University of Chicago Microbiologist Infected from Possible Lab Accident." *New Science.* http://www.sciencemag.org/news/2011/09/updated-university-chicago-microbiologist-infected-possible-lab-accident.

Kane, Paul. (October 17, 2006) "Luck May Have Spared Aides in Anthrax Attack." *Roll Call.*

Klotz, Lynn C., and Edward J. Sylvester. (2009) *Breeding Bio Insecurity.* Chicago: University of Chicago Press.

Knickerbocker, Brad. (August 29, 2006) "In Iraq, Fewer Killed, More Are Wounded." *Christian Science Monitor.*

Koblentz, Gregory D. (2009) *Living Weapons: Biological Warfare and International Security.* New York: Cornell University Press.

Korbitz, Mark. (December 31, 2010) Comments on draft manuscript.

Kortepeter, Mark G., and Gerald W. Parker. (July–August 1999) "Potential Biological Weapons Threats." *Emerging Infectious Diseases* 5 (4): 523–527.

Kotch, Kelly. (August 1, 2010) "Human Performance Optimization: Maximizing the Capability of Our Warfighters." Newsroom, Force Health Protection and Readiness, Office of the Secretary of Defense.

Krueger, Albert P., et al. (1962) "Studies on the Effects of Gaseous Ions on Plant Growth." *Journal of General Physiology* 45:879–895. http://www.ncbi.nlm.nih.gov/pmc/articles/PMC2195229/pdf/879.pdf.

Kurzweil, Ray. (November 30, 2005) "An Exponentially Expanding Future from Exponentially Shrinking Technologies." Council on Foreign Relations lecture, New York. http://www.cfr.org/technology-and-foreign-policy/exponentially-expanding-future-exponentially-shrinking-technology/p9334.

Lakoff, Andrew, and Stephen J. Collier, eds. (2008) *Biosecurity Interventions.* New York: Columbia University Press.

Lamothe, Dan. (October 27, 2014) "New Obama Plan Calls for Implanted Computer Chips to Help U.S. Troops Heal." *Washington Post.*

Leitenberg, Milton, and Raymond A. Zilinskas. (2012) *The Soviet Biological Weapons Program: A History.* Cambridge, MA: Harvard University Press.

Lele, Ajey (2009) *Strategic Technologies for the Military: Breaking New Frontiers.* London: SAGE.

Lewer, Nick, ed. (2002) *The Future of Non-lethal Weapons: Technologies, Operations, Ethics and Law.* Portland, OR: Frank Cass.

Lin, Patrick, Maxwell J. Mehlman, and Keith Abney. (January 1, 2013) *Enhanced Warfighters: Risk, Ethics, and Policy.* New York: Greenwall Foundation.

Little, Robert. (November 19, 2006) "Dangerous Remedy: Military Doctors in Iraq Say That Factor VII Saves Wounded Soldiers, but Other Doctors and Medical Research Suggest That It Can Cause Fatal Clots." *Baltimore Sun.*

Lockett, John. (September 11, 2012) "Army Human Systems S&T." U.S. Army Research, Development and Engineering Command. http://www.ndia.org/Divisions/Divisions/HumanSystems/Documents/John%20Lockett.pdf.

Lockwood, Jeffrey A. (2010) *Six-Legged Soldiers: Using Insects as Weapons of War.* New York: Oxford University Press.

Machiavelli, Niccolò. (1521) *The Art of War.* http://www.feedbooks.com.

Mack, Andrew. (January 1975) "Why Big Nations Lose Small Wars: The Politics of Asymmetric Conflict." *World Politics* 27 (2): 175–200.

Marble, Sanders. (2010) "Why the Military Makes Public Health a Priority." Foreign Policy Research Institute, *Footnotes* 15 (4).

Markon, Jerry. (February 27, 2015) "Internal Audit Slams DHS for Canceling Technology to Fight Bio-Threats." *Washington Post.*

Massingham, Eve. (2012) "Conflict without Casualties . . . a Note of Caution: Non-lethal Weapons and International Humanitarian Law." *International Review of the Committee of the Red Cross* 94 (886): 673–686.

Maurer, Stephen M., ed. (2009) *WMD Terrorism: Science and Policy Choices.* Cambridge, MA: MIT Press.

Maxygen Corporation. (September 28, 1999) "Maxygen Announces $6.7 Million Grant from DARPA to Develop Aerosolized Vaccines." http://www.prnewswire.com/news-releases/maxygen-announces-67-million-grant-from-darpa-to-develop-aerosolized-vaccines-74542327.html.

McLeary, Paul. (June 3, 2015) "The Pentagon Anthrax Scandal Is Getting Worse by the Day."*Foreign Policy.*

Medical News. (May 16, 2007) "Research Reveals Airborne Viruses Can Spread Far and Wide." http://www.news-medical.net/news/2007/05/16/25192.aspx.

Melson, Ashley. (2003) *Bioterrorism, Biodefense, and Biotechnology in the Military: A Comparative Analysis of Legal and Ethical Issues in the Research, Development, and Use of Biotechnological Products on American and British Soldiers.* Berkeley Electronic Press. http://law.bepress.com/expresso/eps/50.

Meredith, L. S., et al. (2007) *Trust Influences Response to Public Health Messages during a Bioterrorist Event.* Santa Monica, CA: RAND.

Michael, Katina. (October 3, 2012) "Meet Boston Dynamics' LS3—The Latest Robotic War Machine." *The Conversation.* http://theconversation.com/meet-boston-dynamics-ls3-the-latest-robotic-war-machine-9754.

Miranda, Robbin A., William D. Casebeer, Amy M. Hein, Jack W. Judy, Eric P. Krotkov, Tracy L. Laabs, Justin E. Manzo, Kent G. Pankratz, Gill A. Pratt, Justin C. Sanchez, Douglas J. Weber, Tracey L. Wheeler, and Geoffrey S. F. Ling. (2015) "DARPA-Funded Efforts in the Development of Novel Brain–Computer Interface Technologies." *Journal of Neuroscience Methods* 244:52–67.

Mooallem, Jon. (February 14, 2010) "Do-It-Yourself Genetic Engineering." *New York Times Magazine.*

Moreno, Jonathan D. (2006) *Mind Wars: Brain Research and National Defense.* New York: Dana Press.

Morrison, Grant. (2011) *Supergods.* New York: Spiegel & Grau.

Mott, Maryann. (January 25, 2005) "Human-Animal Hybrids Spark Controversy." *National Geographic.* http://news.nationalgeographic.com/news/2005/01/0125_050125_chimeras.html.

Munoz, Daniel. (October 11, 2011) "Ukraine Brings Back Naval Killer Dolphins." Reuters.

Murphy, Kim. (April 7, 2012) "A Fog of Drugs and War." *Los Angeles Times.*

Murphy, Sean. (1985) *No Fire, No Thunder.* New York: Monthly Review Press.

Mutsuko, Murakami. (December 18, 1998) "The Cult That Won't Die." *Asia Week.*

National Institutes of Health. (June 11, 2010) *About Recombinant DNA Advisory Committee.* http://oba.od.nih.gov/rdna_rac/rac_about.html.

———. (July 25, 2013) "Silky Brain Implants May Help Stop Spread of Epilepsy." http://www.nih.gov/news/health/jul2013/ninds-25.htm.

National Interagency Biodefense Campus. (2009) "Stories of Service." http://www.detrick.army.mil/nibc/stories.cfm.

National Research Council, Committee on Advances in Technology and the Prevention of Their Application to Next Generation Biowarfare Threats. (2006) *Globalization, Biosecurity, and the Future of the Life Sciences.* Washington, DC: National Academies Press.

———, Committee on Capturing the Full Power of Biomaterials for Military Medical Needs. (2004) *Capturing the Full Power of Biomaterials for Military Medicine: Report of a Workshop.* Washington, DC: National Academies Press.

———, Committee on Opportunities in Biotechnology for Future Army Applications. (2001) *Opportunities in Biotechnology for Future Army Applications.* Washington, DC: National Academies Press.

Nordin, James D., et al. (May 2008) "Bioterrorism Surveillance and Privacy: Intersection of HIPAA, the Common Rule, and Public Health Law." *American Journal of Public Health* 98 (5): 802–807.

Nordland, Max. (February 27, 2013) "20 Afghan Police Officers Killed in 2 Attacks, Including a Mass Poisoning." *New York Times.*

North, Carol S., et al. (August 2005) "Concerns of Capitol Hill Staff Workers after Bioterrorism Focus Group Discussions of Authorities' Response." *Journal of Nervous and Mental Disease* 193 (8): 523–528.

NTI. (November 20, 2002) "Kazakhstan: Vozrozhdeniya Anthrax Burial Sites Destroyed." *Global Security Newswire.* http://www.nti.org/d_newswire/issues/newswires/2002_11_20. html#10.

———. (2003) *China and Chemical and Biological Weapons (CBW) Nonproliferation.* http://www.globalsecurity.org/wmd/world/china/cbw.htm.

———. (2009) *India Biological Chronology.* http://www.nti.org/media/pdfs/india_biological. pdf?_=1321484946.

Nuclear Threat Initiative. (2009) *India Profile: Biological Overview.* http://www.nti.org/learn/ countries/india/biological.

Panen, Sandra, ed. (1985) *Biotechnology: Implications for Public Policy.* Washington, DC: Brookings Institution.

Pash, Chris. (October 8, 2015) "Science Meets Fiction as 3D Printers Create Body Parts and Repair the Unrepairable." *Business Insider.* http://www.businessinsider.com.au/author/ chris-pash.

Pearson, Alan, Marie Chevrier, and Mark Wheelis, eds. (2007) *Incapacitating Biochemical Weapons: Promise or Peril?* Lanham, MD: Rowman & Littlefield.

Petro, James, Theodore R. Plasse, and James A. McNulty. (2003) "Biotechnology: Impact on Biological Warfare and Biodefense" *Biosecurity and Bioterrorism Biodefense Strategy, Practice and Science* 1 (3).

Pierce, Elizabeth. (October 13, 2011) "Anthrax Attack Victims Break Their Silence." *Roll Call.*

Pogrebin, Abigail. (November 1998) "Two Weapons against Terrorism." *Brill's Content* 1 (4).

Pollack, Andrew, and Duff Wilson. (May 27, 2010) "Safety Rules Can't Keep Up with Biotech Industry." *New York Times.*

Preston, Thomas. (2009) *From Lambs to Lions: Future Security Relationships in a World of Biological and Nuclear Weapons.* Lanham, MD: Rowman & Littlefield.

Price, Richard. (1997) *The Chemical Weapons Taboo.* Ithaca, NY: Cornell University Press.

Purdue University. (June 20, 2001a) *Biotechnology Promises Major Advances for U.S. Army.* http://news.uns.purdue.edu/UNS/html4ever/010620.Ladisch.Army.html.

———. (June 20, 2001b) *Future Army Could Run on Alternative Fuels, Photosynthesis.* http:// news.uns.purdue.edu/UNS/html4ever/010620.Ladisch.NRCenergy.html.

RAND. (1994) *The Fly on the Wall and the Jedi Knight.* http://www.rand.org.

Rappert, Brian. (1999) "Assessing Technologies of Political Control." *Journal of Peace Research* 36 (6): 741–750.

———. (2001) "The Distribution and the Resolution of the Ambiguities of Technology; or Why Bobby Can't Spray." *Social Studies of Science* 31 (4): 557–592.

———. (2005) "Prohibitions, Weapons and Controversy: Managing the Problems of Ordering." *Social Studies of Science* 35 (2): 211–240.

Reel, Monte. (August 20, 2006) "Brazil's Road to Energy Independence." *Washington Post.*

Regents of the University of California. (2013) "The Institute for Collaborative Biotechnologies." https://www.icb.ucsb.edu.

Reilly, Ryan J. (November 2, 2011) "Feds: Four Members of Georgia 'Fringe Militia Group' Plotted Biological Attack on Citizens, Government Officials." *TPMMuckraker.* http:// tpmmuckraker.talkingpointsmemo.com/2011/11/feds_four_members_of_georgia_fringe_ militia_group_plotted_biological_attack_on_citizens_government_officials.php?ref=fpblg.

Reissman, Dori B., et al. (2004) "One-Year Health Assessment of Adult Survivors of *Bacillus anthracis* Infection." *Journal of the American Medical Association* 291: 1994–1998.

Reiter, Dan, and Allan C. Stamm. (2002) *Democracies at War.* Princeton, NJ: Princeton University Press.

Religious Tolerance Council. (November 4, 1998) "Aum Shinri Kyo." Ontario, Canada. http://www.religioustolerance.org/dc_aumsh.htm.

Reuters. (November 24, 2012) "WWII Code Found on Long-Dead Pigeon in England May Never Be Broken." *Washington Post.*

Rexton Kan, Paul. (March 2008) "Drug Intoxicated Irregular Fighters: Complications, Dangers, and Responses." Strategic Studies Institute, Army War College. http://www.strategicstudiesinstitute.army.mil/pubs/summary.cfm?q=850.

Rifkin, Jeremy. (1998) *The Biotech Century.* New York: Putnam.

Ritchie, Elspeth Cameron. (June 20, 2013) "A Smoking Pillbox: Evidence That Sgt. Bales May Have Been on Lariam." *Time.*

Rizwan, Sharjeel. (September 2000) "Revolution in Military Affairs." *Defence Journal.* http://www.defencejournal.com/2000/sept/military.htm.

Roberts, Brad, ed. (1993) *Weapons of the Future?* Washington, DC: CSIS.

Roberts, Guy B. (2003) *Arms Control without Arms Control: The Failure of the Biological Weapons Convention Protocol and a New Paradigm for Fighting the Threat of Biological Weapons.* Institute for National Security Studies Occasional Paper No. 49, United States Air Force Academy, Colorado.

Robotics Trends. (April 5, 2010) "Cyclone Power Technologies Delivers Biomass-to-Power System to Robotic Technology." http://www.businesswire.com/news/home/20100330005245/en/Cyclone-Power-Technologies-Delivers-Biomass-to-Power-System-Robotic.

Rothman, Peter. (February 15, 2015) "Biology Is Technology—DARPA Is Back in the Game with a Big Vision and It Is H+." *Humanity+.* http://hplusmagazine.com/2015/02/15/biology-technology-darpa-back-game-big-vision-h.

Sagan, Scott D. (July 8, 2013) "Deterring Rogue Regimes: Rethinking Deterrence Theory and Practice." ASCC Final Report, Stanford University Center for International Security and Cooperation.

Saletan, William. (May 29, 2013) "The War on Sleep." *Slate.* http://www.slate.com/articles/health_and_science/superman/2013/05/sleep_deprivation_in_the_military_modafinil_and_the_arms_race_for_soldiers.html.

Sandberg, Anders. (June 11, 2014) "The Five Biggest Threats to Human Existence." *Washington Post.* http://www.washingtonpost.com/posteverything/wp/2014/06/11/the-five-biggest-threats-to-human-existence/?hpid=z4.

Savulescu, Julian, and Nick Bostrom, eds. (2009) *Human Enhancement.* Oxford: Oxford University Press.

Schreiweis, Christiane, et al. (2014) "Humanized Foxp2 Accelerates Learning by Enhancing Transitions from Declarative to Procedural Performance." *Proceedings of the National Academy of Sciences of the United States of America* 111 (39): 14253–14258.

The Scotsman. (March 3, 2002) "SNLA Threats and Attacks." http://www.scotsman.com/news/scotland/top-stories/snla-threats-and-attacks-1-605332.

Segarra, Alejandro. (2002) *Agro-terrorism: Options for Congress.* Washington, DC: Congressional Research Service.

SelectUSA. (2013) "The Biotechnology Industry in the United States." http://selectusa.commerce.gov/industry-snapshots/pharmaceutical-and-biotech-industries-united-states.html.

Shachtman, Noah. (March 1, 2007) "Be More Than You Can Be." *Wired.*

Shane, Scott. (April 22, 2010) "Colleague Rebuts Idea That Suspect's Lab Made Anthrax in Attacks." *New York Times.*

Shelton, Jeffery. (AY 2004–2005) "Emerging Biotech Regulatory Policy," The Industrial College of the Armed Forces, Student Paper, Fort McNair, Washington DC.

Shenyang Liaoling Ribao. (October 2, 1997) "Biological Effects of Gene Weapons Revealed." http://wnc.fedworld.gov.

Sigger, Jason. (February 9, 2009) "Top Army Biowar Lab Suspends Research after Toxin-Tracking Scare (Updated Again)." *Wired.*

Singer, P. W. (2010) "The Ethics of Killer Applications: Why Is It So Hard to Talk about Morality When It Comes to New Military Technology?" *Journal of Military Ethics* 9 (4): 299–312.

Singh, Gunjan. (September 5, 2008) "Role of Biotechnology in Defence." Institute for Defence Studies and Analyses. http://idsa.in/event/rolebiotecnologyindefense_alele_050908.

Smith, Frank L., III. (2014) *American Biodefense: How Dangerous Ideas about Biological Weapons Shape National Security.* Ithaca, NY: Cornell University Press.

———. (2011) "A Casualty of Kinetic Warfare: Military Research, Development, and Acquisition for Biodefense." *Security Studies* 20 (4): 663–696.

Smithson, Amy. (2011) *Germ Gambits: The Bioweapons Dilemma, Iraq, and Beyond.* Palo Alto, CA: Stanford University Press.

Soh, Hyongsok. (2013) "Synthetic Viral Mimics for Programmed Drug Delivery." Institute for Collaborative Biotechnologies. https://www.icb.ucsb.edu/research/biotechnology-tools/synthetic-viral-mimics-programmed-drug-delivery.

Sokolove, Michael. (January 7, 2007) "The Scold." *New York Times.*

Specter, Michael. (May 25, 2010) "A Life of Its Own." *New Yorker.*

Spiegel, Alix. (April 22, 2010) "When the 'Trust Hormone' Is Out of Balance." *All Things Considered*, National Public Radio. http://www.npr.org/templates/story/story.php?storyId=126141922.

SteelFisher, Gillian K., et al. (2012) "Public Response to an Anthrax Attack: A Multiethnic Perspective." *Biosecurity and Bioterrorism: Biodefense Strategy, Practice, and Science* 10 (4): 401–411.

Stern, Jessica. (2000). *The Ultimate Terrorists.* Cambridge, MA: Harvard University Press.

———. (Winter 2002–2003) "Dreaded Risks and the Control of Biological Weapons." *International Security* 27 (3): 89–123.

Sun Tzu. (Lionel Giles, translator, 2009) *The Art of War.* El Paso, TX: El Paso Norte Press.

Sunshine Project. (April 6, 2004) *US State Department Launches New Push to Use Agent Green in Colombia.* http://www.sunshine-project.org/publications/pr/pr060404.html.

———. (February 1, 2008) *Introduction to Biological Weapons.* http://www.sunshine-project.org/bwintro.

Sylvester, Edward, and Lynn C. Klotz. (1987) *The Gene Age.* New York: Scribner.

Temple-Ralston, Dina. (October 10, 2015) "The Secretive Government Agency Where 'Anything Imagined Can Be Tried.'" *Washington Post.*

Texas Department of State Health Services. (2007) *History of Bioterrorism.* http://www.dshs.state.tx.us/preparedness/bt_public_history.shtm.

Tharoor, Ishaan. (February 13, 2009) "Why Chemical Warfare Is Ancient History." *Time.* http://www.time.com/time/world/article/0,8599,1879350,00.html.

Tian, Jingdong, and Ishtiaq Saaem. (December 13, 2007) "Light-Controlled Smart Material Exploits Billions of Years of Evolutionary Performance Tuning." *Nanowerk.* http://www.nanowerk.com/spotlight/spotid=3674.php.

Todhunter, Colin. (January 21, 2013) "Genetically Engineered 'Terminator Seeds': Death and Destruction of Agriculture." *Global Research.* http://www.globalresearch.ca/genetically-engineered-terminator-seeds-death-and-destruction-of-agriculture/5319797.

Trader, Tiffany. (February 19, 2015) "Army Research Lab Lays Out HPC Strategy." *HPCWire.* http://www.hpcwire.com/2015/02/19/army-research-lab-lays-out-hpc-strategy.

Truth Commission. (2001) *Special Report.* Matter AM8079/97. http://sabctrc.saha.org.za/documents/decisions/59536.htm.

Tu, Anthony T. (2002) *Chemical Terrorism: Horrors in Tokyo Subway and Matsumoto City.* Fort Collins, CO: Alaken Inc.

Tucker, Jonathan B., ed. (2000) *Toxic Terror: Assessing Terrorist Use of Chemical and Biological Weapons.* Cambridge, MA: MIT Press.

United Nations. (1992) *Convention on Biological Diversity.* New York: United Nations Press. http://www.cbd.int/convention/text/default.shtml.

United States Arms Control and Disarmament Agency. (2015) *Adherence to and Compliance with Arms Control Agreements.* http://www.state.gov/t/avc/rls/rpt/2015/243224.htm#China.
———. (2005) *Adherence to and Compliance with Arms Control Agreements.* http://www.state.gov/t/avc/rls/rpt/51977.htm.
United States Army. (1998) *Analysis and Control of Polymer Interphases in Fibers and Films.* http://www-sscom.army.mil/services/BIOTECH.
———. (1998) *New Materials Development Using Biotechnology Process.* http://www-sscom.army.mil/services/BIOTECH.
———. (1998) *Sustained Food Quality.* http://www-sscom.army.mil/prodprog/sustain.
United States Department of Defense. *Anthrax as a Biological Warfare Agent.* DefenseLink. http://www.defenselink.mil.
———. *DoD Biological Warfare Threat Analysis.* DefenseLink. http://www.defenselink.mil.
———. (1998) *Technical Annex.* DefenseLink. http://www.defenselink.mil/cgi-bin.
United States Department of Justice. (2010) *Amerithrax Investigative Summary.* https://www.justice.gov/archive/amerithrax/docs/amx-investigative-summary.pdf.
United States Department of the Army. (1956) *The Law of Land Warfare* (FM 27–10). http://armypubs.army.mil/doctrine/DR_pubs/dr_a/pdf/fm27_10.pdf.
United States Food and Drug Administration. (2015) *Cellular and Gene Therapy Products.* http://www.fda.gov/biologicsbloodvaccines/cellulargenetherapyproducts/default.htm.
United States Government Accountability Office. (July 31, 2008) *United States Postal Service: Information on the Irradiation of Federal Mail in the Washington, D.C., Area.* GAO-08-938R. http://www.gao.gov/products/GAO-08-938R.
———. (September 2008) *Biosafety Laboratories: Perimeter Security Assessment of the Nation's Five BSL-4 Laboratories.* http://www.gao.gov/new.items/d081092.pdf.
United States House of Representatives Subcommittee on Prevention of Nuclear and Biological Attack. (July 13, 2005) *Engineering Bio-Terror Agents: Lessons from the Offensive U.S. and Russian Biological Weapons Programs.* Washington, DC: Government Printing Office.
United States Navy. (2003) "U.S. Navy Marine Mammal Program." http://www.spawar.navy.mil/sandiego/technology/mammals/index.html.
University College of London. (April 28, 2008) *Results of World's First Gene Therapy for Inherited Blindness Show Sight Improvement.* http://www.ucl.ac.uk/media/library/Genetherapyblind.
Van Atta, Richard. (2008) "Fifty Years of Innovation and Discovery" in *DARPA: 50 Years of Bridging the Gap.* Arlington, VA: Defense Advanced Research Projects Agency.
Vastag, Brian, and David Brown (February 1, 2012) "Recommendation to Censor Bird Flu Research Driven by Fears of Terrorism." *Washington Post.*
Vaughn, Elaine, et al. (2012) "Predicting Response to Reassurances and Uncertainties in Bioterrorism Communications for Urban Populations in New York and California." *Biosecurity and Bioterrorism: Biodefense Strategy, Practice, and Science* 10 (2): 188–204.
Vijayaraghavan, K., and Yeoung-Sang Yun. (May–June 2008) "Bacterial Biosorbents and Biosorption." *Biotechnology Advances* 26 (3): 266–291.
Viotti, Paul R., Michael A. Opheim, and Nicholas Bowen. (2008) *Terrorism and Homeland Security.* Boca Raton, FL: CRC Press.
Vogel, Kathleen. (2012) *Phantom Menace or Looming Danger? A New Framework for Assessing Bioweapons Threats.* Baltimore, MD: Johns Hopkins University Press.
W. L. Gore & Associates. (2014) "Selectively Permeable Fabric." http://www.goreprotectivefabrics.com/remote/Satellite/GORE-CHEMPAK-Fabrics/GORE-CHEMPAK-Selectively-Permeable-Fabric.
Wade, Nicholas. (January 14, 1998) "Researchers Extend Lifespan of Human Cells Far beyond Normal Limit." *New York Times.*
Walker, Hunter. (April 29, 2013) "Everett Dutschke Makes First Court Appearance in Ricin Case." *Talking Points Memo.* http://tpmdc.talkingpointsmemo.com/2013/04/everett-dutschke-makes-first-court-appearance-in-ricin-case.php.
Walkin, Malham, ed. (1986) *War, Morality, and the Military Profession.* Boulder, CO: Westview Press.

Walzer, Michael. (1977) *Just and Unjust Wars: A Moral Argument with Historical Illustrations*. New York: Basic Books.

Wang, Lei. (June 16, 1998) "Development of Biological Detecting Technology by the U.S. Military." *Beijing Renmin Junyi* [People's Military Surgeon] 41 (2). http://wnc.fedworld.gov.

Wang, Songjun. (June 16, 1998) "Scientific and Technological Advances Related to Biological Weapons and Their Defense." *Beijing Renmin Junyi* [People's Military Surgeon] 41 (2). http://wnc.fedworld.gov.

Warrick, Joby. (September 5, 2013) "As Syria Deteriorates, Neighbors Fear Bioweapons Threat." *New York Times*.

Warwick, Graham. (May 22, 2009) "DARPA Plans Triple-Target Missile Demo." *Aviation Week*. http://www.aviationweek.com/aw/generic/story_channel.jsp?channel=defense&id=newTRIPLE052209.xml.

Watson, Julie. (December 17, 2015) "VA Sets National Policy for Robotic Legs for Paralyzed Veterans." Associated Press.

Weatherford, Jack. (2004) *Genghis Khan and the Making of the Modern World*. New York: Three Rivers Press.

Whalen, Jeanne. (May 12, 2009) "In Attics and Closets, 'Biohackers' Discover Their Inner Frankenstein." *Wall Street Journal*.

Wheelis, Mark, Lajos Rozsa, and Malcolm Dando, eds. (2006) *Deadly Cultures: Biological Weapons since 1945*. Cambridge, MA: Harvard University Press.

Whitlark, Rachel, and Amir Stepak. (February 2010) *Unconventional Ties? States, Non-State Actors, and Weapons of Mass Destruction*. Presented at the Annual Conference of the International Studies Association.

Williams, Reggie. (March 14, 2003) "Brentwood Workers Fear Returning after Anthrax." *Washington Afro-American*. http://www.highbeam.com/doc/1P1-79665992.html.

Willman, David (June 10, 2011) "Inside Our Own Labs, the Threat of Another Anthrax Attack." *Washington Post*.

———. (August 23, 2012) "Early Warnings on Biowatch." *Los Angeles Times*.

Wright, Susan, ed. (2002) *Biological Warfare and Disarmament: New Problems/Perspectives*. Lanham, MD: Rowman & Littlefield.

Xu, Rebecca. (April 15, 2013) "XNA: A New Genetic Language." *Dartmouth Undergraduate Journal of Science*. http://dujs.dartmouth.edu/biological_sciences/xna-a-new-genetic-language#.U5Z6Y_m1bYg.

Yong, Ed. (April 19, 2012) "Synthetic XNA Molecules Can Evolve and Store Genetic Information, Just Like DNA." *Discover*. http://blogs.discovermagazine.com/notrocketscience/2012/04/19/synthetic-xna-molecules-can-evolve-and-store-genetic-information-just-like-dna/#.UcPct_k3CSo.

Yorktown Technologies LP. (2010) "GloFish: Frequently Asked Questions." http://www.glofish.com/faq.asp.

Zaits, Les. (April 14, 2011) "Rajneeshees' Utopian Dreams Collapse as Talks Turn to Murder." *Oregonian*. http://www.oregonlive.com/rajneesh/index.ssf/2011/04/part_five_utopian_dreams_die_i.html.

Zhang, Sarah. (February 11, 2015) "Watch This Man Scale a Rock Wall with DARPA's Bionic Arm." *Gizmodo*. http://gizmodo.com/watch-this-man-scale-a-rock-wall-with-darpas-bionic-arm-1685253239.

Zigmond, Jessica. (July 21, 2008) "Restructuring BioShield." *Modern Healthcare*. http://www.modernhealthcare.com/article/20080721/REG/39338593.

Index

About the Author

David Malet teaches international relations at the University of Melbourne. Previously he was assistant professor of political science and director of the Center for the Study of Homeland Security at Colorado State University–Pueblo. From 2000 to 2003 he served as a national security aide to US Senate Majority Leader Tom Daschle. His research includes leading an experimental study of risk communications after biological attacks that was funded by the US Environmental Protection Agency. He also analyzed transnational militant recruitment in his first book, *Foreign Fighters*, and he has advised several governments and international organizations on this issue.